*The man who can think and does not know how to express what he thinks is at the level of him who cannot think.*

Pericles

# EFFECTIVE BUSINESS AND TECHNICAL PRESENTATIONS SECOND EDITION

## MANAGING YOUR PRESENTATIONS BY OBJECTIVES AND RESULTS

**GEORGE L. MORRISEY**

President, MOR Associates
*Buena Park, California*

**ADDISON-WESLEY PUBLISHING COMPANY**
Reading, Massachusetts • Menlo Park, California
Don Mills, Ontario • Wokingham, England • Amsterdam
Sydney • Singapore • Tokyo • Mexico City • Bogotá
Santiago • San Juan

*Fifteenth Printing, April 1985*

ISBN 0-201-04828-0
    OPQ-AL-898765

# PREFACE TO SECOND EDITION

**WHY A "SECOND EDITION?"**

There are two fundamental reasons. First, and most obvious, is the need for an update. The book came out originally in 1968 and was a product of ideas and materials that were developed as early as 1962. There have been some changes in the state of the art (although, interestingly enough, not nearly as many as one might think). Also, there has been quite a bit of feedback from users of the first edition that indicates a need for a certain amount of "fine tuning." This is further supported by my own experiences in conducting seminars and in observing and practicing the art of effective business and technical presentations.

Second, I want to tie it closer conceptually to *Management by Objectives and Results* (MOR), the approach to management with which I have become identified. Interestingly, and to prove a point I make in my MOR seminars, I was practicing the approach long before I knew what it was. Perhaps that is one of the reasons why the original training program and the first edition of this book were so successful. The idea of *establishing objectives for the briefing* and then putting it together in such a way that they are accomplished is such a logical, commonsense approach that results-oriented people were naturally attracted to it. From there, the transition to MOR which, from my admittedly biased point of view, is the only way to effectively manage anything, is equally natural.

## WHAT'S DIFFERENT ABOUT THIS EDITION?

Conceptually, there is nothing different. The basic approach introduced in the first edition has stood the test of time. The differences are more in degree than in kind. Consequently, organizations that have been using the first edition will have no difficulty in making the transition to this one. Here are the principal differences:

1. *Elimination/reduction of male dominance.* Through the good-natured badgering of several of my feminist colleagues (notably Mary Fuller, Dru Scott, Doris Seward and Theo Wells), I have consciously worked to eliminate male-dominant references. While I do not agree with some of the approaches advocated within the feminist movement, I must, in all good conscience, acknowledge the right of women to be looked upon and referred to as equals. If I have slipped on any such references in the new edition, it is a result of long-entrenched habit patterns and not by intent.

2. *MOR reference.* A new sub-title, modified and additional copy in the text, and an Appendix article have been inserted to show the natural relationship of this particular communications skill to the broader concept of *Management by Objectives and Results* (MOR).

3. *New worksheets.* These have been added for use in preparing a Preliminary Plan, Resource Material Selection, and planning the Presentation. They are also included on perforated pages in the Appendix, along with others, where they may be removed and reproduced for continued use.

4. *Governmental references and illustrations.* In support of the wide use the first edition has received in the public sector, several references to and illustrations of governmental briefings have been added.

5. *Audio Visual Aids expansion.* There are enough new developments in this area to justify a complete book in itself. I have attempted to highlight those tools and

techniques that I feel will be of most use to the average presentor.

6. *Preliminary Arrangements expansion*. This portion of the chapter on Making the Presentation has been expanded to include guidelines for various seating arrangements and a reproducible Preliminary Arrangements Checklist.

7. *Brief annotated bibliography.*

8. *Appendix Articles*. In addition to the one on MOR identified above, there are two articles written especially for this book by experts in their fields. The first, *Storyboarding. . . For Briefings* by Gus Matzorkis identifies a unique approach to getting meaningful involvement from others in generating and assembling ideas. The second, *Videorecording. . . For Briefings* by Tom Sechrest deals with various applications of this relatively new communications medium.

9. *Instructor's Kit*. Available separately from the publisher, this includes suggested lesson plans, visuals, an audio tape, worksheets, etc., all that is needed to set up your own training program on Effective Presentations.

## ACKNOWLEDGMENTS

My appreciation, as always, goes first of all to the many participants in my seminars who have forced me to continue working on improving the approach. In addition, I have received constructive feedback from many individuals and organizations who have used the first edition of the book, identifying "soft" and confusing areas that needed expansion or fine tuning. I am particularly grateful to: Gus Matzorkis, not only for his fine article on *Storyboarding* but for being a good friend and counselor at the right time; Reed Royalty of Pacific Telephone, who attended one of my early public seminars on Effective Presentations and developed one of the most comprehensive in-company training programs on the subject I have seen, which has influenced many of the modifications included in the new edition and in the Instructor's Kit; Jack Rush of Rockwell International, my former training partner, for his continued constructive feedback on this and others of my management development efforts; and to Tom Sechrest, a young but rapidly rising star in the field of communications, for his excellent article on *Videorecording*.

*Buena Park, California*                                      G.L.M.
*October 1974*

# PREFACE TO FIRST EDITION

**THIS BOOK IS USELESS**

unless you approach it with the idea, that, with careful thought and adequate preparation, an effective oral presentation or briefing can be made on any subject. This is not a collection of shortcuts and gimmicks that will make you a polished presentor. Nor is it a substitute for participation in a formal training program, although it can serve as an excellent text for such a program. It *will* provide:

1. A tested step-by-step method that will result in a concise, interesting, and effective presentation.
2. Practice exercises, in connection with each of the steps, that you can do with the assistance of others, to ensure effective application of the suggested techniques.
3. Guidelines for a wise selection and use of audio-visual aids.
4. Instruction on the pitfalls you must avoid.
5. Insights for improving the communication relationship with your audience.
6. Techniques for increasing the effective use of your body and your voice.
7. Suggestions on making the best use of audience question periods.

## WHY WAS THIS BOOK WRITTEN?

Actually, this book is the translation of a training program in Briefing Techniques which I first developed in 1962 at North American Aviation's Space Division in Downey, California. At that time, line management recognized the growing requirement for oral presentation of ideas in what has come to be known as the industrial briefing.

Such presentations were required from a wide variety of key management and technical personnel, most of whom had had little or no training or experience in making such presentations. It became increasingly apparent that a significant number of work-hours were being expended in the preparation, presentation and, even more critically, in listening to many briefings that were neither brief nor effective in conveying their message. Therefore, we were charged with the responsibility for developing a training program to remedy this situation.

The purpose of this program was not to develop polished public speakers. Many academic and commercial programs are already effectively doing this. Rather, its purpose was to provide the average manager or technical expert with a set of tools and guides for practicing so that he or she could give a *brief*, coherent, and reasonably successful oral presentation.

In the nearly five years since the course was developed, more than 1400 people at North American have completed the program under several different instructors. Furthermore, there is a continuing waiting list of employees who want to participate in it *on their own time*. While there are no specific statistics on the subject, periodic post-class surveys of participants have indicated the following general results:

1. Reduction of from 20% to 50% in *briefing preparation time*.

2. Reduction of more than 50% in *presentation time* for the same type of briefing (before the program, average presentation time for many participants was an hour or more; after the program, average time for the same participants was 20 to 30 minutes).

3. Substantial reduction, without loss of effectiveness, in the number and complexity of charts used.

4. Significantly increased self-confidence about giving briefings.

5. Much better reception by, and corresponding effectiveness with, the people to whom the briefings were given.

6. Confirmation of these results by the superiors of many of these participants.

Several former participants in the program, as well as my fellow instructors, urged me to put the course into manual or textbook form. A survey of the field revealed that there were many publications on the subject of public speaking. However, there appeared to be none specifically designed as a "how to" for the relatively inexperienced person in industry, government, or business who must make a largely technical presentation to a critical audience. For these reasons this book was written. .

Needless to say, experience in conducting this training program at North American and elsewhere resulted in many modifications and an increasingly effective approach. I believe that this book captures the essence of the program in its current form. It can be most productive as a text in a formal training program with a skilled instructor. If no such instructor is available, the next best approach would be for a group to work together for mutual improvement. Having a skilled briefer in to lead the critiques and practice sessions would make the training more effective. However, if these suggestions are not practicable, an individual can use this text effectively as a self-teaching device. It is designed to be a practical, down-to-earth guide that will be used and re-used regularly. (Note that the high points of the text are grouped together in the Appendix, which is perforated for greater convenience of the user.)

## WHO CAN BENEFIT FROM THIS BOOK?

As mentioned earlier, this material is designed primarily for the relatively inexperienced presenter in an industry, government, or business, not for the public speaker as such (although that person too could benefit from the ideas presented here). Typical of those who would find this a valuable guide are:

*President* of a company, for a report to the Board of Directors or stockholders;

*Sales Engineer,* for a technical sales presentation to customer representatives;

*Controller,* for an overview of the company's financial projections to a top management group;

*Manufacturing Cost Analyst,* for a review of staff loading requirements with the department manager;

*Research Scientist,* for presentation of the results of a study

   a.  at a formal gathering of peers (e.g., a national symposium),

   b.  to management people not oriented to that technical field;

*Industrial Engineer,* for a review of a work-sampling study to the management of a less-than-fully-cooperative department;

*Credit Manager,* for introduction of a new credit-application system to employees;

*Governmental Department Head,* for presentation of an annual budget forecast to the appropriate legislative body;

*Training Instructor,* for presentation of a training lecture;

*Employment Specialist,* for an employee recruitment presentation;

*Project Engineer,* for a report on the current status of a directed design change

   a.  to his or her own management,

   b.  to the customer;

*Purchasing Agent,* for a bid-seeking meeting with potential subcontractors;

*Supervisor,* for a motivational presentation to subordinates on workmanship;

*Safety Representative,* for an accident-prevention presentation to a group of maintenance supervisors;

*Anyone,* with a requirement to present primarily technical information in a brief and understandable manner to a critical audience.

**HOW CAN THIS BOOK BE USED IN A TRAINING PROGRAM?**
(See Effective Presentations Instruction Kit for further descriptive material.)

The personal preference and experience of the instructor and the particular circumstances of the moment will have considerable bearing on the approach to conducting the program. My own experience has proved successful when the following points were observed:

1. *Optimum group size,* 12 to 15 persons

    I have conducted fruitful classes, however, with as few as 7 and as many as 25, in the latter case using a second instructor for divided practice sessions.

2. *Optimum program length,* 21 to 27 hours

    It could easily be expanded to a full semester in a school situation, with greater subject depth and with more and longer practice presentations with the following options:

    a. weekly sessions of 2, 3, or 4 hours (3 usually seems best),

    b. one 2- to 3-day intensive seminar-workshop,

    c. combination of half- and full-day sessions.

*Note:* program length is directly related to the number of participants because of the requirement for individual practice presentations.

3. *Preparation exercises*

    a. Pre-assignment—come to first session with a briefing topic (and general knowledge of subject matter) in mind.

      b. Write briefing objectives in class with small group critique, instructor circulating.

      c. Prepare Preliminary Plan, in or out of class, with in-class small group critique, instructor circulating.

      d. When practical and desirable, design charts, in or out of class, for instructor and group critique.

4. *Practice presentations by participants*

      a. At least 2 presentations of 10- to 15-minute duration (some variation up or down in duration is possible without loss of value).

      b. Tape recording (audio or video) for later self-critique by the presenter.

      c. Written and verbal critique by instructor and fellow participants.

5. *Instructor approach*

      a. Demonstrate variety of aids and approaches during formal presentations.

      b. Schedule practice presentations to start as quickly as possible after preparation material and related exercises have been covered, making initial assignments during first session.

      c. Discuss techniques related to "Making the Presentation" in short increments, interspersed as a change of pace between groups of practice presentations.

Again, let me emphasize that these considerations have proved effective for me in conducting these programs. Another instructor may be equally successful using a different combination. The material can be readily adapted to almost any approach.

## HOW CAN THIS BOOK BE USED BY AN INDIVIDUAL?

Recognizing that many individuals will want or, of necessity, have to use this book without benefit of an accompanying training program, I suggest the following approach for maximum benefit. As with most tools, the versatility and usefulness of this book will increase in direct proportion to the individual's effort and experience in using it. It will be of most value to you if you:

1. Skim through it quickly to get an overview.
2. Then, read it carefully, doing the recommended practice exercises.
3. Use it as a specific guide every time you make an oral presentation.
4. Refer to it for solving specific problems only after you are familiar with the total recommended concept.
5. Practice the recommended techniques every chance you get.
6. Start now!

## ACKNOWLEDGMENTS

There are many individuals to whom I am indebted for much of the material that is developed here. Over a period of several years the hundreds of participants in my classes on this subject have really been the most help in refining and solidifying the approach. The feedback they have given me on the continued benefits they gained from this approach has provided me with the incentive to write this book.

Among many present and former co-workers who have made contributions to the content of this text, I want to express particular appreciation to:

Albie Johnson, for the substantial help he gave me in developing the material on audio-visual aids in the original training program;

Dave Lewis, my boss, for his continued encouragement as well as for his critical content review;

Gene Phillips, who served as an excellent Devil's Advocate on the manuscript draft and provided some specific ideas in the area of audience question periods;

Tom Scobel, for his critical suggestions and encouragement during my first cut at writing it; and to

John Traband, for his suggestions during the development of the original training program and particularly for his ideas on the use of the Audience Retention Curve.

Finally, I want to offer my apologies as well as my thanks to Carol, Lynn, and Steve, my wife and children, for putting up with me during the hectic periods of writing and rewriting.

*Downey, California*                                                    G.L.M.
*August 1967*

# CONTENTS

## 4 MAKING THE PRESENTATION

## 5 CONCLUSION

## BIBLIOGRAPHY

## APPENDIX A

## APPENDIX B    (removable worksheets)

# INTRODUCTION

## WHAT'S IN IT FOR YOU?

How often have you sat through a business or technical presentation which

- went way over your head;
- was too elementary for you and the people around you;
- appeared to have little purpose other than to satisfy the need of the presentor to expound;
- used visuals that, even if they could be seen adequately, served more to confuse than clarify;
- seemed to go on and on with no conclusion reached;
- was an obvious ordeal for both the presentor and and the audience;
- could have been handled more effectively and efficiently through some other form of communication?

If we have not struck a familiar chord with some or all of these, your experience in participating in such presentations has been extremely limited.

If your reaction from being on the receiving end of such a presentation is somewhat less than highly enthusiastic, how would you rate yourself when you are on the transmitting end? So many managers and technical experts, who are extremely competent in their own fields, completely ignore the same kinds of analytical processes that made them experts in the first place when trying to convey their ideas to others. While, clearly, some presenters will be more effective than others, you *will* increase your own presentation effectiveness, regardless of its current level, by following the relatively simple analytical approach described here. However, while it may be "simple," that does not mean it will be "easy." It will require a discipline that may be somewhat foreign to your customary practice in the past. However, the potential payoff in satisfaction, and accomplishment will be tremendous.

## HOW DOES THIS RELATE TO MOR?

MBO or, as I prefer to call it, MOR (*Management by Objectives and Results*)* is one of the most widely accepted approaches to management in use today. It is a process that focuses on output or accomplishment rather than on input or activity. As such, it moves from the general to the specific, subdividing an effort that is large and complex until it reaches a unit size that is manageable. (See Appendix A for a brief overview of the process.)

"Managing" a presentation is no different from managing any other kind of investment. In allocating certain resources, primarily ideas, time, and energy in this case, you expect to get a return that will exceed the value of the investment. In other words, you anticipate that your presentation, if successful, will bring about a result that has value for you. Defining that anticipated result in the form of an objective and designing the presentation to specifically accomplish that is what this process is all about.

While this is an approach to using a technical skill, it is also management in its purest sense. Thus, the MOR process can be applied to anything that has meaning for you whether it be your job, your career, your life, or, in this case, a presentation that you must make.

---

*See *Management by Objectives and Results* by George L. Morrisey (Reading, Mass.: Addison-Wesley, 1970).

## WHAT DO WE MEAN BY BRIEFING?

The term *briefing* will be used interchangeably with *oral presentation* throughout this text. For our purposes, briefing may be defined as *the preparation and presentation of critical subject matter in a logical and condensed form, leading to effective communication.* Its essential components are illustrated in the accompanying figure.

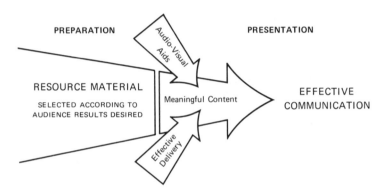

Briefing is made up of two major parts: *preparation* and *presentation*. In reality, there should be no firm line between the two. Only a rare individual can properly present a briefing entirely prepared by someone else. The person who will make the actual presentation ought at least to provide direction in terms of concepts and approaches to be used so that the briefing will reflect his own convictions and personality. Many an otherwise well-prepared briefing has gone "down the tubes" because it was given by someone who did not have a part in its preparation. We will have more on this later.

**Audience results desired** is the determining factor in selecting the resource material from the variety of sources available. This factor is a vital one, and must be held paramount in the mind of the person preparing the briefing.

**Meaningful content** is at the heart of the presentation. Without it, you may have a speech, but you won't have a briefing. The presentation is *supported by* effective delivery and audio-visual aids. Understanding the *support* role of these techniques is critical. If the content is compromised in order to exploit unusual or available aids or gimmicks, or a specific delivery approach, there is a strong risk that you will win the battle but lose the war. You may gain immediate acceptance of the presentation but not accomplish its real objectives.

**Effective communication** should be the goal. Simply defined, this means getting the message across in a manner that will accomplish its objectives. If your audience does what you want it to do as a result of your briefing, you will have achieved effective communication, even though you have violated every principle covered in this text!

The three arrows in the diagram at the beginning of this chapter illustrating *meaningful content, audio-visual aids,* and *effective delivery* identify the principal parts of this text. Meaningful content, of course, is the heart of the presentation. The following chapter is devoted to "Steps in Preparation of a Briefing." Each of these steps is critical to the effective determination of the briefing content. Overlooking any one of them leads to serious flaws in the presentation material. For this reason Chapter 2 takes up the largest portion of the text and deserves the greatest attention.

"Selection and Use of Audio-Visual Aids," covered in Chapter 3, really has a place in both the preparation and the presentation portions. Audio-visual aids play such a critical part in most briefings that they deserve special treatment. Even though their role is only to support the presentation, the use of wrong or poorly designed aids can destroy an otherwise good briefing. The rules and suggestions covered in that chapter can substantially improve the effectiveness of any presentation.

Prime factors to be considered in making the actual presentation are covered in Chapter 4. Three of the four major subheadings in the chapter, *Preliminary Arrangements, Platform Techniques,* and *Vocal Techniques,* can apply to virtually any speaking situation. The fourth, *Audience Question Techniques,* is somewhat peculiar to the briefing presentation. The effective application of the points discussed will lend substantial credence to the subject matter and increase the interest and attention of the audience.

## TYPES OF BRIEFINGS

Four major types of briefings are in common practice in business, industry, and government. While there is considerable overlap, and there may be elements of all four in a particular briefing, we can usually identify them as follows:

1. **Persuasive** (some of which is in every briefing). First and foremost, of course, you must sell your audience on the fact that you know what you are talking about. However, more specifically a persuasive briefing might be used to:

   a. Excite the interest of a potential customer (we will use the word "customer" as synonymous with "user" since anyone who uses your product or service is, and rightfully should be considered, a "customer") or group of customers in a new product, service, or capability you are offering or will offer.

   b. Cause the audience to have confidence in the organization you represent and the message you are presenting.

   c. Sell an existing customer on a new product, modification of present product or method, change in scope or funding, schedule delays, or a new procedure.

   d. Sell upper management on the need for additional personnel, money or methods.

   e. Sell co-workers, subordinates, or members of parallel organizations on changes in operations or on the need for working together more effectively.

2. **Explanatory.** This approach is usually designed to provide a general familiarization, give the big picture, or identify new developments. The presenter's primary objective is

not trying to sell anything, even though the element of persuasiveness is essential, but rather to make available to the audience new or renewed knowledge and understanding. It is a broad-brush approach and rarely should involve heavy detail. It might be used for such purposes as:

a. Orienting new employees to the organization.

b. Acquainting staff members with what is involved in opening a new branch or division of the organization.

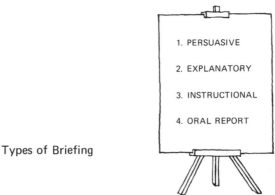

Types of Briefing

1. PERSUASIVE

2. EXPLANATORY

3. INSTRUCTIONAL

4. ORAL REPORT

c. Orienting members of other departments, companies, or agencies who could benefit from a general knowledge of the subject matter.

d. Presenting information to service clubs, civic organizations, or other groups for good public relations.

3. **Instructional.** The purpose of this briefing is to teach others how to use something, such as a new procedure or piece of hardware. This objective usually requires greater involvement of the audience to ensure proper learning and is the most frequent type of briefing where heavy detail is required. Typical uses might be for:

a. In-plant instruction to customers' employees on the use of services you are supplying.

b.  Instruction to customers' representatives on how to instruct their own employees on the use of equipment.

c.  Instruction to management, co-workers, or subordinates on the use of procedures.

d.  Instruction to other departments or organizations on the effective melding of their operations with yours.

4. **Oral Report.** This briefing normally brings the audience up to date on something with which they are already familiar. It may or may not involve heavy detail on a selective basis, according to the needs and interests of the specific audience. The briefing may be a report to:

a.  Customer representatives on current progress in terms of meeting their requirements.

b.  Management on money and personnel expenditures against budget.

c.  Co-workers or subordinates on current schedule progress and possible trouble spots.

## SUMMARY

The effectiveness of your presentation will be determined to a large extent by the degree to which the steps to proper preparation and presentation have been covered. These steps are not a panacea that will solve all your briefing problems, but they do provide a logical and proven approach which should result in shorter and more productive presentations. Thus, if you manage your presentation with primary emphasis on *objectives and results*, the payoff on your investment can be tremendous.

# STEPS IN THE PREPARATION
# OF A BRIEFING

## INTRODUCTION

As with any important job, a good briefing requires careful planning. Inadequate planning results in considerable wheel spinning and, not infrequently, a poorly organized briefing that is too long, too detailed, or ineffectual. Careful thought during the preparation period may cause us to realize that the required communication can be accomplished more efficiently and productively by some other means (for example, by a letter, memo, bulletin, telephone call, or motion picture), and that the briefing is not actually needed. A briefing is expensive, primarily in terms of the *time* investment of everyone concerned, including the audience, and it should not be held unless it is truly the most effective method of communication. The way you speak is and should be less formal than the way you would write. Careful thought, though, can eliminate fuzzy impressions and make for a crisper briefing.

The secret to an effective presentation is to break down its preparation into manageable units, starting with the results or response you want to achieve. In other words, we apply the MOR process to our presentation. Systematic following of the six steps in preparation will not work miracles. However, it is almost certain to result in your spending less time on preparation, producing a shorter, clearer briefing, and eliminating unnecessary briefings. Also, the likelihood that your ideas will be accepted by the audience will be much greater. Each of

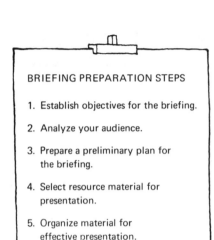

BRIEFING PREPARATION STEPS

1. Establish objectives for the briefing.

2. Analyze your audience.

3. Prepare a preliminary plan for
   the briefing.

4. Select resource material for
   presentation.

5. Organize material for
   effective presentation.

6. Practice the presentation in advance.

Briefing Presentation Steps

these steps plays a vital role in the preparation process and none should be overlooked. We will take up each of these steps, one at a time, devoting a separate section to each. Let us first look at them all together.

You will note as you examine each of the preparation steps from number 2 on that parts of previous steps are incorporated as a base from which the new one is developed. This apparent redundancy demonstrates and emphasizes the cumulative aspect of each successive step. The establishment of objectives, for example, becomes a basic consideration in each of the following preparation steps.

Some of these steps are followed intuitively by many briefers. This discussion is designed to give specific definition to the steps, with an illustrated breakdown of their practical application.

**SELECTION OF BRIEFING TOPIC
FOR PRACTICE EXERCISES**

For most effective learning in how to use these preparation steps, *select a briefing topic* that you might be called on to present in the near future. The more "real world" you can make it, the better your learning will be. Using that topic throughout, do the exercises suggested with each step. Here are some suggestions for possible briefing topics, in the event you do not have one in mind.

1. Sales presentation to a prospective customer/client.

2. Request for support from upper management, for example, to secure a budget increase for a particular project.

3. Presentation of functions of your department.

4. Recruiting presentation on the merits of joining your organization to a group of prospective employees or members.

5. Description of the functions of your own job to

   a. fellow employees,

   b. your superior,

   c. your family,

   d. members of a civic organization to which you belong,

   e. a group of high school students.

6. Instruction to an employee or a group of employees on how to perform a specific job you are assigning to them.

7. Report to your superior on the current status of a major task assignment for which you have responsibility.

Your selection of a topic need not be restricted to these. The more practical you can make the topic selection in terms of your own specific requirements, the more valuable the exercises will be. Once your topic is selected, follow these steps in order.

BRIEFING PREPARATION STEPS

1. **Establish objectives for the briefing.**

2. Analyze your audience.

3. Prepare a preliminary plan for the briefing.

4. Select resource material for presentation.

5. Organize material for effective presentation.

6. Practice the presentation in advance.

Briefing Presentation: Step 1

## STEP 1: ESTABLISH OBJECTIVES FOR THE BRIEFING

Without doubt, this is the most critical step in preparation, and the one which is most often overlooked. We usually are so concerned about *what* we are going to say that we lose sight of *why* we are saying it. In other words, the first question we should ask ourselves is

"Why am I giving this briefing?"

If there isn't a satisfactory answer, perhaps the briefing doesn't need to be given at this time or to this group of people. Evidence indicates that a significant number of briefings in many organizations are unnecessary. They serve no useful purpose and are a poor use of time for those involved in giving the briefing as well as for those required to attend.

In this step, we are concerned with *short-term* results. What will be the measure of a successful presentation: approval of my plan, an awareness of a need leading to the initiation of corrective action, understanding of and ability to use properly the information being presented, creation of a favorable image leading to future acceptance of ideas?

There is a natural tendency to confuse the objectives of what we are proposing in the briefing with the objectives of the briefing itself. THEY ARE NOT THE SAME! For example, in a new product proposal to top management, an objective for the *project* might be "to get additional profitable business for the company." However, the objective for the *briefing* would be "to get approval to proceed with the project." An objective for a new education *program* in the Department of Health, Education, and Welfare might be "to provide citizens from disadvantaged areas with more saleable skills." An objective of a *briefing* within the Department probably would be "to initiate action on research and development" or "to get a budgetary allocation for Phase 1." It is critical that the difference between these two kinds of objectives is clearly understood. The first stated objective in each case is a laudable long-range achievement and, undoubtedly, would be a valuable benefit to identify in the briefing itself. However, the briefing cannot accomplish that. It can only set it in motion. The briefer must recognize what the briefing *will* accomplish, if successful, and must focus his or her efforts to that end. If the short-term objective of the *briefing* is not accomplished, there is very little likelihood that the long-term objective of the *project* will be accomplished.

Clearcut objectives give a firm direction to the briefing preparation. Instead of following the normal initial impulse to start wading through reports, studies, charts, and so forth, much of which will be tossed aside later, you should invest your time at the outset to determine the objectives. Doing so will eliminate much wasted effort and pin-point the target toward which all material should be aimed. Your concern should be first with the answer to the basic question "Why?" not "What?"—"Why am I giving this briefing" and *not* "What am I going to put into it?" These objectives are, essentially, self-centered. What results do *I* want to get from this briefing? The objectives will not, necessarily, be shared directly with the audience.

## Realistic Definition of Results Expected

It is important to keep in mind the specific *results* expected from the briefing, and they should be:

1. **Realistic in scope**, so they can be accomplished in both the preparation time and the presentation time available. It is far better to do an adequate job of presenting one major step of a program, opening the door for later presentations, than to try to cover the waterfront, so to speak, and run the risk of overwhelming the audience or doing a poor job on some of the steps. The most fantastic and enjoyable meal I ever had was one that took six hours to complete. Each course was prepared and presented at just the right time for maximum satisfaction: time for digestion was allowed between courses. The master chef exacted a comfortable profit on each meal, and he had a waiting list of customers that required reservations a year or more in advance. Presenting a briefing in well-prepared, digestible segments often is the key to acceptance of the ideas.

2. **Realistic in view of the audience's knowledge and background.** Does the audience have the knowledge and background necessary to achieve the results you want? Presenting the applications of a new computer in programming terms might be unrealistic for a group of production executives. A preliminary briefing on terminology might be in order or perhaps a broad, results-oriented presentation, with provision for later coverage of programming, after they feel knowledgeable about the

subject and interested in its application to their fields of interest.

3. **Realistic in view of the audience's ability to act.** Do the audience members have the authority to make the decisions you may wish? For example, if you are presenting a briefing on the "Need for Increased Training in the Company" to a group of first-line supervisors, it would be unrealistic to propose increasing the present training staff or adding facilities or equipment. They would not be able to make such a decision. However, proposing that they acquire greater familiarity with the capabilities of the Training Department, that they take time to train themselves, and provide training time for subordinates, or that they give both verbal and written support to training efforts would be realistic.

4. **Realistic in terms of what you can reasonably expect to accomplish.** The gold-plated-Cadillac solution may be the best one in a given situation. If, however, such things as budget, personnel, floor space, or unconquerable resistance by key individuals will make acceptance of this idea unrealistic, it might be better to set your objective now for an achievable stripped-down Chevy, hoping that later on it can be traded in for a better model.

## Criteria for Judging Objectives of a Briefing

The objectives for a briefing should be specific and they should meet *one or more* of the following criteria. They need not be in complete sentences as illustrated, but they should incorporate all critical elements.

1. They should answer the question, "Why am I giving this briefing?"

   Positive answer: *This is a regular monthly briefing to keep upper management informed on the current status of the*

*XYZ project and to draw their attention to any potential trouble spots so that corrective action can be initiated.*

am I giving this briefing?

2. They should state the *results* desired from the briefing, in effect completing the sentence, "I want the following things to happen as a result of this briefing . . ."

   Positive answer: *The customer will recognize a need for additional units of my product and will agree to purchase them.*

Getting the Desired Results

3. If it is important to identify the body of knowledge to be presented, this objective should be qualified in terms of results expected—"I want to tell about . . . *so that* . . . will take place."

   Positive answer: *I want to tell my subordinates about the new work process so that they will recognize its value to them in increased efficiency and the fact that it will not represent a threat to their jobs.* (Note the difference between this objective and a "How to do it" objective.)

**Hidden objectives.** Since this is, legitimately, a self-centered step which you may or may not choose to share with your audience, you may wish to include any hidden or indirect objectives you hope the briefing will accomplish. The purpose in including them would be as a reminder for you or anyone helping you in the briefing preparation. Examples of hidden objectives:

- To make the audience aware of and responsive to my other products/services/capabilities.
- To create a favorable receptivity for future presentations, regardless of the outcome of this one.
- To relieve anxieties over possible production cutbacks.
- To gain my boss's approval of my ability to communicate to top management.

Using the "Need for Increased Training in the Company" as a briefing topic, objectives of the presentation might be:

1. To create an *awareness of the need* for increased training.
2. To gain *management approval and support* for increased training so that they will act to
   a. authorize necessary funds,
   b. authorize time for training, and
   c. give verbal and written support to training efforts.

For a briefing on a "New Educational Program for Disadvantaged Areas," the following might be appropriate objectives:

1. To reach agreement on need for new program.
2. To initiate action on research and development
3. To get a budgetary allocation for Phase 1.

For a briefing on "Support Role of Financial Department," try this:

1. To explain the support functions of the Financial Department to line managers so they will
   a. recognize and use those services available to them,

    b. accept inputs from Financial Department staff as desirable and helpful,

    c. seek assistance before rather than after problems occur, and

    d. provide positive feedback for service improvement.

## PRACTICE EXERCISES

1. Using the topic you selected (a practical one on which you might give a briefing in the near future), write a short clear statement of its objectives in line with the criteria we have outlined under Step 1.

2. When you are reasonably satisfied with your statement, ask a few of your co-workers, or class or group members, to critique it on its own merits, in other words, with no comment or explanation from you. If they understand exactly what you are trying to accomplish with the briefing, after reading the statement just once, you have written it well. If they agree that the objectives are realistic in terms of the criteria established above, you have accomplished what is probably the most important and perhaps the most difficult step in the preparation. If they have objections which seem valid to you, rewrite the statement of your objectives in line with their suggestions, modifying them in line with your own good judgment.

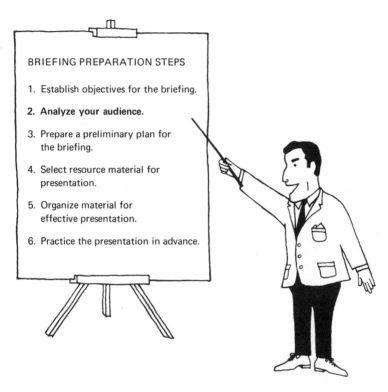

Briefing Presentation: Step 2

## STEP 2: ANALYZE YOUR AUDIENCE

In taking the first step, establishing objectives, you will be primarily self-centered, determining what you, the briefer, want to accomplish. The second step, audience analysis, though in many ways an integral part of setting objectives, reverses the point of view. Now you are required to consider the audience to whom the briefing will be presented. The most successful briefings are those that are prepared with a particular audience in mind, and tailored to suit the knowledge, attitudes, likes and dislikes of the members of that audience.

Many an otherwise well-prepared briefing has fallen short of its objectives because the briefer failed to anticipate how the audience would react. For instance, one organization whose members were to give regular briefings to a top-level group of government officials, prepared what were basically very good presentations. For the first four months, however, they experienced a very cool reception from this group of officials. They could not understand this because the material was what the officials had requested. Furthermore, they had prepared an excellent set of charts to illustrate the major points, using cartoon figures in some cases to build interest, a technique that can be very effective. *However,* in this case, using cartoons was unfortunate because the key member of this group of officials had a personal dislike for the cartoon-figure approach. This personal idiosyncracy and his attitude affected the response of the others. A short preliminary outline of the presentation or a brief discussion of the format with one or two of the officials at the planning stage might have prevented this trouble. In any event, after the first briefing was something less than successful, a real attempt should have been made to determine why.

The information here is primarily for your assistance in preparing the briefing in advance. Recognize also that it may be necessary to do some audience analysis "on the spot" and make some modifications in your briefing based on data that was not available to you in advance. However, the more thorough job of audience analysis you have done in advance, the less "on the spot" adjustment you should have to make. Also, there would be many audiences for whom a detailed analysis as described here would be totally unnecessary. Whether you use the recommended guide or not, you must give attention to your audience's viewpoint before preparing your briefing.

## Audience Analysis Audit (AAA)

This guide (see pages 23 and 24) may not guarantee success, but it is an effective tool for giving you a picture of your audience and helping you determine how best to angle your briefing. You

Who are they?

may also want to investigate other characteristics of your audience that are not identified in this guide. The AAA is divided under four subheadings, each of which provides information for a specific use:

1. *Identifying objectives for this audience* was discussed under Step 1. Getting a firm picture of what you want to accomplish will point up many of the specific audience characteristics to look for.

2. *Specific analysis of this audience* is designed to help you determine the *scope of the material*. It should indicate how deep into the subject it is advisable to go *from the point of view of the audience*.

3. *General analysis of this audience* should provide insight into the kind of overall *approach* most likely to achieve the objectives with this particular audience.

4. *Information and techniques* concentrates on the *type of format* and *the specific approaches* that would most likely have a positive effect on the specific audience. (Further study in this area might have prevented the cartoon fiasco.)

The important point is to take a systematic approach in analyzing your audience so that you not only avoid as many problems as possible, but also turn out the briefing most likely to accomplish your objectives *with this audience*. If you get into trouble with your briefing, it might be because you failed to pay enough attention to your AAA. You can get this advance information about your audience in a variety of ways:

1. Inquiring of other persons who have briefed this audience.
2. Conducting a postmortem or debriefing after each presentation to assess audience reaction.
3. Reviewing reports from or about members of this audience.
4. Inquiring directly or indirectly from selected members of the audience or others associated with them.
5. Thinking the situation out logically and applying common sense to what you already know about the situation and the audience.

**PRACTICE EXERCISES**

1. Using the *Audience Analysis Audit* (AAA) as a guide, and supplying whatever additional information about the audience you feel would be helpful, prepare a profile or general description of the audience for whom you are preparing the particular briefing you selected earlier in this chapter.
2. Review your analysis or observations with a few of your co-workers to check on whether they would agree that your profile is as accurate as possible.
3. Identify in writing (for inclusion in your Preliminary Plan) the specific audience for whom you are designing this briefing and give a one- or two-sentence summary of *pertinent* information about their knowledge, attitudes, and so forth.

## AUDIENCE ANALYSIS AUDIT (AAA)*

(Fill in the blanks or circle the terms most descriptive)

---

1. Identify the objectives in presenting your briefing to THIS audience. What do you want to happen as a result of it?

   _____

   _____

   (Keep these objectives in mind as you consider the items below.)

2. **Specific Analysis** of members of this audience —

   a. Their knowledge of the subject:

   High level      General    Limited    None    Unknown

   b. Their opinions about the subject and/or the speaker or organization represented:

   Very favorable          Favorable              Neutral
   Slightly hostile        Very Hostile           Unknown

   c. Their reasons for attending this briefing:

   _____

   d. Advantages and disadvantages of briefing objectives to them as individuals:

   Advantages _____

   Disadvantages _____

3. **General Analysis** of members of this audience —

   a. Their occupational relationships to the speaker or organization:

   Customer    Top management    Immediate management
   Peers       Subordinates      Other management
   Other workers    Public

---

*For your convenience, a copy of *Audience Analysis Audit* will be found at the back of this book.

## Audience Analysis Audit (AAA) (continued)

---

  b. Length of relationship with organization as customer or employee:

   New  Less than two years  More than two years
   Unknown

  c. Their vocabulary understanding level:

   Technical     Nontechnical  Generally high
   Generally low    Unknown

  d. Open-mindedness (willingness to accept ideas to be presented)

   Eager      Open     Neutral
   Slightly resistant  Strongly resistant  Unknown

4. Information and Techniques —

  a. Information and techniques most likely to gain the attention of this audience:

   Highly technical information  Statistical comparisons
   Cost figures  Anecdotes  Demonstrations  Other

   _____

  b. Information or techniques likely to get negative reactions from this audience:

   _____

5. Summarize, in a few sentences, the most important information from the preceding four sections.

   _____

   _____

   _____

   _____

   _____

   _____

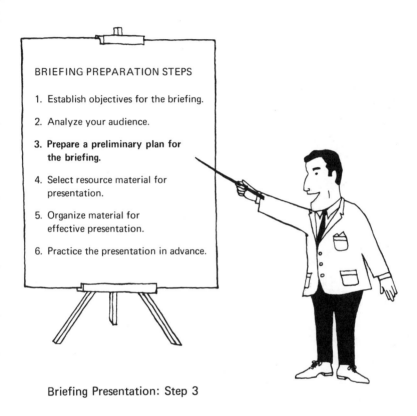

BRIEFING PREPARATION STEPS

1. Establish objectives for the briefing.

2. Analyze your audience.

3. **Prepare a preliminary plan for the briefing.**

4. Select resource material for presentation.

5. Organize material for effective presentation.

6. Practice the presentation in advance.

Briefing Presentation: Step 3

## STEP 3: PREPARE A PRELIMINARY PLAN FOR THE BRIEFING

There is still another step to follow *before* you decide exactly what subject matter to include in your briefing. The Preliminary Plan is like a blueprint. Its purpose is to build a framework on which a briefing can be developed and to help you decide how much and what kind of material you will need. The Preliminary Plan is not designed to be a speaking outline, but a conceptual approach to what will most logically lead to accomplishing your objectives.

## How to Prepare a Preliminary Plan

The guidelines and accompanying worksheets which appear on the following pages are perhaps the most useful part of this entire text. Individuals who are frequently required to prepare briefings often keep a copy of these guidelines under glass on their desks or in some other prominent spot where they can readily refer to it. (For your convience a copy of these guidelines appears on perforated sheets at the back of this book so that you can remove and use them.) One of the *extra* values gained from the effective use of these guidelines is the ability to prepare a briefing on extremely short notice, even in as little as twenty minutes.

Items 1 and 2, which have already been discussed, serve as the base for the accompanying guidelines. Item 3 relates to the objectives and calls for stating the *main ideas* or *concepts* which the audience *must get* if the objectives of the briefing are to be met. This is really the heart of the plan.

The *main ideas* should be statements in *conclusion form,* conclusions you want the audience to reach about the material presented. These statements should *not* merely identify subjects to be covered, such as organization, cost, schedule; *they should spell out what you want the audience to believe about them*. These may or may not be statements of fact. However, they should provide guidance for the kinds of facts to be presented. For example:

1. The organization has the capability of performing as required.
2. The cost is reasonable and will be more than offset by future value received.
3. A realistic schedule can be maintained if we have effective management support.

Frequently there may be only one main idea in a briefing, even though it may be approached from several angles. You may wish to stress only the idea that "We will meet our primary objectives in spite of certain temporary setbacks."

## GUIDELINES FOR PREPARING A PRELIMINARY PLAN*

---

1. Identify specific objectives for the briefing, keeping in mind the following criteria:
   a. They should answer the question, "Why am I giving this briefing?"
   b. They should state the results desired from the briefing, in effect, completing the sentence, "I want the following things to happen as a result of this briefing: . . ."
   c. They should be designed to accomplish whatever hidden objectives you have for the briefing.

Note: If the body of knowledge to be presented must be identified in the objectives, use a sentence such as "I want to tell about . . . so that . . . will take place."

2. Identify the specific audience for whom you are designing this briefing and state in a one- or two-sentence summary pertinent information about their knowledge, attitudes, and so forth.

---

*For your convenience, a copy of *Guidelines for Preparing a Preliminary Plan* will be found at the back of this book.

## Guidelines for Preparing a Preliminary Plan (continued)

---

3. State the MAIN IDEAS OR CONCEPTS that the audience MUST get if the objectives of the briefing are to be met. These should:
   a. Be in conclusion form and preferably in complete sentences.
   b. Definitely lead to the accomplishment of the specific objectives.
   c. Be interesting in themselves or capable of being made so.
   d. Be few in number, usually no more than five.

4. Identify under each main idea the types of factual information necessary so that this audience can understand these ideas. Avoid excessive detail.

## This Plan Should be Used as a Guide:

1. For the briefer in selecting materials, keeping ideas channeled, and determining emphasis points.

2. For support personnel who may provide the backup data, prepare charts and other aids, and assist in the briefing itself.

Only rarely should you try to present more than five main ideas if you want them to come through clearly and be remembered by your audience. If it is necessary to get across a larger number of ideas in order to achieve some specific objective, perhaps it would be better to present them at two meetings, the first an overview or a request for input, and the second a bid for decision-making. For example, consider the following.

### SAMPLE TWO-SESSION APPROACH

The following might be main ideas for two briefings related to presenting a plan for computerizing certain company operations.

### First Briefing

1. Many required statistical reports can be prepared with greater speed, greater accuracy, and at a lower overall cost through the use of computers.
2. The company is evaluating potential usage to determine the feasibility of in-house versus contracted computer services.
3. There will be no reduction of personnel as a result of computerization.
4. Based on specific guidelines (provided), recommendations for additional computer usage must be received by (date).

### Second Briefing

1. Decision has been made to contract for outside computer service for the first six months.
2. Company A has been awarded the contract on the basis of greater probable overall value to us.

3. Requests for computer service will be coordinated by the XYZ department in an equitable and expeditious manner.

4. Each department head should continually evaluate his or her operations with a view to determining how effectively the computer services are being used.

5. The computer service will be evaluated at the end of the first five months to determine whether or not we want to renew the contract or take some other action.

Again, the main ideas, as stated in the Preliminary Plan, may or may not be facts in themselves. They do not have to be. They must be conclusions that can be reached, however, on the basis of the factual information presented in the briefing.

Item 4 of *Guidelines* requires identifying the *types* of factual information *necessary* to support and clarify the main ideas. Detail should be kept to a minimum unless the purpose of the briefing is to instruct or to supply specified detailed information in some general area with which the audience is already familiar. Using the briefings on computerization described above, we might find the following types of supporting factual information appropriate:

## First Briefing, Supporting Information

Idea 1

a. Sample reports scheduled for computerization.

b. Flow chart showing input and return.

c. Samples of typical accuracy problems.

d. Comparative cost figures.

Idea 2

a. Estimated in-house cost figures for staff, facilities, equipment, supplies, and so forth.

b. Estimated contract service cost figures, including in-house co-ordination expense.

c. Volume requirements and quality-cost figures for economic use of current and proposed approaches.

Idea 3

a. Areas of service expansion to be implemented.

b. Retraining plans required.

Idea 4

a. Typical areas for expanded computer usage.

b. Guidelines for determining efficient computer usage.

## Second Briefing, Supporting Information

Idea 1 Reasons for decision determination.

Idea 2 Reasons for selection of Company A.

Idea 3

a. Reasons for coordination by the designated department.

b. Process of requesting computer services.

Idea 4 Suggested evaluation procedures—key points to identify.

Idea 5

a. Methods of evaluation.

b. Key determinants for subsequent decisions.

As pointed out earlier, the Preliminary Plan may be the
most useful step in the entire approach to the briefing. This plan
has two basic functions:

1. Forcing the briefer to determine carefully the direction to
take, serving as a guide in selecting the subject matter,
keeping the flow of ideas channeled, and indicating where
emphasis can best be placed.

2. Serving as a guide for support personnel who provide the
back-up data, prepare charts and other aids, or who assist
in the actual presentation. Many briefers have had the ex-

perience of assigning such responsibility to someone and having the work done all wrong in terms of what the briefer had in mind. The Preliminary Plan, if prepared properly, minimizes this problem because it gives a specific *written* basis from which the briefer and those in a supporting role can work.

Which direction shall I take?

The person giving the briefing should be actively involved in preparing the Preliminary Plan, regardless of his or her level in the organization or the amount of time available for the briefing. Only in this way will it reflect the briefer's point of view, interest, knowledge, and ability. If he or she lacks sufficient time, interest, or knowledge to participate effectively in the preparation of the Preliminary Plan, a serious question should be raised as to whether or not it might be better for someone else to make the presentation.

**PRACTICE EXERCISES (See Sample Preliminary Plans)**

1. Write a Preliminary Plan, following the *Guidelines for Preparing a Preliminary Plan,* for the briefing topic you selected. Use the *Preliminary Plan Worksheet,* starting on the next page, as an aid in this process. (A copy which may be reproduced will be found at the back of this book.) Concentrate particularly on Item 3, the main ideas or concepts. Be certain that they are stated in the form of the *conclusions* you want the audience to reach, and are not merely a list of topics. These statements should in most cases be complete sentences.

2. Ask some of your co-workers to critique your Preliminary Plan, particularly the main ideas. Find out whether the statements are clear to them; ask what suggestions they may have which would clarify the meaning. See whether they agree that the main ideas meet the criteria outlined under *Guidelines.* Find out whether they feel that these ideas will accomplish your objectives. Determine also whether they approve of your choice of factual information. Ask for any suggestions they may have.

Do not feel you must justify your own approach. Their immediate reactions may be very helpful to you, but you are the one who has been chosen to give this briefing, who has studied and given careful thought to the content, and who should have enough self-confidence to accept criticism without being defensive.

## PRELIMINARY PLAN WORKSHEET*

Title or subject of this briefing: _____

_____

Approximate date, time, and place of this briefing: _____

_____

Who requested the briefing (if other than yourself)? _____

_____

Your **OBJECTIVES** for _this_ briefing (what will be the immediate results if this briefing is successful?):

1. _____

_____

2. _____

_____

3. _____

_____

4. _____

_____

**AUDIENCE** for this briefing (who are they and what is their general knowledge of, interest in, and attitude toward the subject?):

_____

_____

_____

---

*For your convenience, a copy of _Preliminary Plan Worksheet_ will be found at the back of this book.

**Preliminary Plan Worksheet (continued)**

---

**MAIN IDEAS OR CONCEPTS** that the audience *must get and retain* if the objectives of the briefing are to be met:

1. _____

   _____

2. _____

   _____

3. _____

   _____

4. _____

   _____

5. _____

   _____

Types of **FACTUAL INFORMATION** necessary to support the main ideas:

Idea 1

_____

_____

_____

_____

Idea 2

_____

_____

_____

_____

## Preliminary Plan Worksheet (continued)

Idea 3

_____

_____

_____

_____

Idea 4

_____

_____

_____

_____

Idea 5

_____

_____

_____

_____

**PRELIMINARY PLAN: Sample 1**

---

**Topic:** New Educational Program for Disadvantaged Areas in Region A

**Objectives:**

1. To reach agreement on need for new program.
2. To initiate action on research and development.
3. To get a budgetary allocation for Phase 1.

**Audience:**

Agency head, functional directors, other regional managers, staff advisers. They are knowledgeable but will tend to be either disinterested in or skeptical of any new programs in field.

**Main ideas** the audience MUST get:

1. Current education programs are not meeting the need for increasing saleable skills in identified areas.
2. New data suggests Program G will produce better results in a shorter time.
3. Phase 1, a pilot project in X community, will furnish valuable research data for broader application while providing citizens there with tangible short-term benefits.
4. Funds for Phase 1 can be made available by eliminating Program D and postponing action on Program E, without negative consequences.
5. This would be a politically popular program "on the Hill."

**Factual supporting information:**

Ideas 1 & 2

   a. Statistics on employment, unemployment, placement.
   b. Survey results from potential employers.
   c. Projected results from Program G.

Idea 3

   a. Rationale for pilot project in X community.
   b. Projected research data anticipated.
   c. Projected tangible short-term benefits for citizens there.

Idea 4

a. Rationale for reallocation of funds.

Idea 5

a. Quotes from Congressional Record and the press.

b. Expressions of interest and support from congressional representative and leaders from X community.

## PRELIMINARY PLAN: Sample 2

---

**Topic:** Support Role of Financial Department

**Objective:**

To explain the support functions of the Financial Department to line managers so they will:

a. Recognize and use those services available to them.

b. Accept inputs from Financial Department staff as desirable and helpful.

c. Seek assistance before rather than after problems occur.

d. Provide positive feedback for service improvement.

**Audience:**

Line managers, by Division. Most will have limited knowledge of or interest in Financial Department functions and will be neutral to slightly hostile.

**Main ideas** the audience MUST get:

1. Financial Department is in business to help make the line manager's job easier and less complicated.

2. Financial Department has many useful services available to the line manager in addition to the accepted maintenance operations (payroll, financial reports, etc.).

3. Financial Department personnel are skilled professionals with a desire to help.

4. Early identification of potential problems can provide more useful and less costly assistance for the line manager.

5. Financial Department is continually seeking feedback from line managers on ways to improve service to them.

**Factual supporting information:**

Ideas 1 & 2

> Brief identification of *only* those services of direct concern to *this* audience.

Idea 3

a. Specific examples of services that have been or could be provided—in terms of benefit to the user (line manager).

b. Brief identification of key staff members likely to relate to *this* audience.

Idea 4

> Specific examples of positive consequences of early identification (avoid or minimize negative consequences of failure to do so).

Idea 5

> Specific examples of improvements made as a result of positive feedback.

**PRELIMINARY PLAN: Sample 3**

---

**Topic:** Progress Report on XYZ Project

**Objective:**

To keep upper management informed in a regular monthly briefing on the current status of the XYZ Project and to draw their attention to any potential trouble spots so that corrective action can be initiated.

**Audience:**

Department Manager, Section Director, Chief Engineer, related staff personnel. They are familiar with the project and will be interested primarily in adherence to schedule.

**Main ideas** the audience MUST get:

1. Project is currently one week behind but can be brought back on schedule with the following adjustments:

    a. Authorization of 100 overtime hours in next month.

    b. Elimination of second reinforcement test which will serve merely to validate the results of the first test.

2. Costs will remain within budget if both these adjustments are made.

3. All performance standards are being met.

**Factual supporting information** related to all the main ideas:

1. Factors affecting schedule.

2. Comparative cost figures.

3. Test results.

4. Key performance measurements.


**PRELIMINARY PLAN: Sample 4**

---

**Topic:** Need for Increased Training in the Company

**Objectives:**

1. To create an awareness of the need for increased training.

2. To gain management approval and support for increased training so they will act to:

    a. Authorize necessary funds,

    b. Authorize time for training, and

    c. Give verbal and written support to training efforts.

**Audience:**

Members of top management plus other management personnel at Director level or higher. Most will have a general knowledge of the subject; a few will be favorably inclined, but most will be neutral, skeptical, or slightly hostile.

**Main ideas** the audience MUST get:

1. Increased training is essential if we are to survive in the industry.

2. Money invested in training now (charged to Overhead or taken from Profit) will be returned manyfold in the future.

3. Time spent in training now (taken from urgent current work) will result in a much more profitable use of time in the future.

**Factual supporting information:**

Idea 1

a. New technology requirements.

b. Training experience in other similar companies.

c. Potential application of new management concepts.

Idea 2

a. Recent training progress in the company.

b. Comparative cost of operation figures (before and after).

c. Personnel training versus replacement costs.

Idea 3

a. Comparative (before and after) time-investment ratios.

b. Intangible time benefits, for example, increased confidence and effectivity of personnel resulting in more productive use of time.

**PRELIMINARY PLAN: Sample 5**

---

**Topic:** Orientation of New Employees

**Objectives:**

1. To provide an overview of the company's history, organization, and products so that new employees can more readily understand and identify with the company's philosophy and objectives and feel a sense of responsibility for the part they will play.

2. To identify, in broad terms, company policies and benefits so that the employees will be able to apply or take advantage of them with a minimum of wasted time and effort.

**Audience:**

New employees at all levels. Most will have at best a limited knowledge of the subject matter, will be somewhat nervous and unsure of themselves, will be unable to retain the majority of the detail presented, but will be positively motivated to learn.

**Main ideas** the audience MUST get:

1. This is a strong company with a good product line and a firm future for competent, conscientious employees.
2. Company management personnel are interested in the growth and well-being of their employees.
3. Capable assistance is available to all employees in the application of policies and procedures and in the interpretation and use of company benefits.
4. Employees are expected to meet their company responsibilities and, in return, will be treated fairly and equitably.

**Factual supporting information** related to all the main ideas:

1. Company historical highlights, major product lines, and growth projections.
2. Company organization and key management personnel.
3. Highlights of relevant company policies and procedures.
4. Company benefit programs and sources of information.

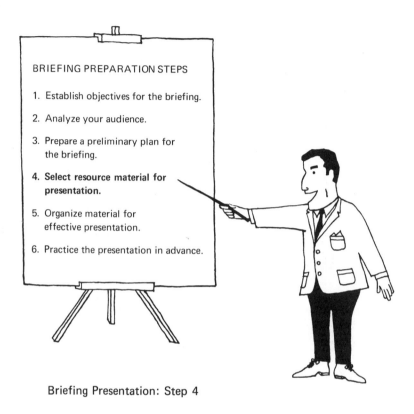

BRIEFING PREPARATION STEPS

1. Establish objectives for the briefing.

2. Analyze your audience.

3. Prepare a preliminary plan for the briefing.

4. **Select resource material for presentation.**

5. Organize material for effective presentation.

6. Practice the presentation in advance.

Briefing Presentation: Step 4

## STEP 4: SELECT RESOURCE MATERIAL FOR THE PRESENTATION

For most briefings, finding enough resource material to include is not a problem. The problem, rather, is one of proper selection, the determination of what and how much available material should be included in the briefing.

While there is no magic formula to guarantee proper selection, asking ourselves some basic common-sense questions as we evaluate the data may be helpful. The proper preparation of a Preliminary Plan is basic, of course, to proper selection.

## QUESTIONS FOR GUIDANCE

These questions will follow the Preliminary Plan to a large degree. See the worksheet at the end of this section for specific application.

1.  What is the **object or purpose** of the briefing?

    *Is it to be a persuasive, explanatory, instructional, or oral report type of briefing? Do you want to arouse interest, test an idea, recommend action, inform, or resolve problems? (Review the objectives in your Preliminary Plan.)*

2.  What should be **covered?** What can best be **eliminated?**

    *Supporting facts identified in the Preliminary Plan should indicate the subject matter to be covered. (For example, in Sample 4: new technology requirements, training experience in similar companies, and so forth.) Items, even though related to the topic, should be eliminated unless they contribute significantly to the accomplishment of the objectives. (For example, in Sample 4: Educational Reimbursement Plan and Employee Performance Review discussion would not aid in reaching the specific objectives, even though both may be a part of the training function.)*

3.  What amount of **detail** is necessary?

    *This depends, of course, upon many factors; preparation and presentation time, the audience and its particular interests, how much it is necessary for them to know for the objectives to be reached. Most briefings we have observed have much more detail than is necessary. It is far better to leave the audience a bit hungry, wanting more detail, than to give them so much that they get confused or bored. You may wish to have the detailed information available in the event you are asked about it, although you do not use it in the actual briefing. We will discuss this matter further when we come to Audience Question Techniques.*

4. What **must** be said if the objectives are to be reached?

   *The answer to this question depends on what you have selected as the main ideas from the Preliminary Plan. You must decide what specific resource material is essential if the main ideas are to be accepted by the audience.*

5. What is the **best** way to say it?

   *Primarily considering the audience, what types of subject matter and what method of presentation (examples, anecdotes, statistics, comparisons) do you feel will be most effective in getting the main ideas across?*

6. What kind of **audience action or response** is required if the objectives are to be met?

   *Do we need to force an immediate response (for example, approval of plan, authorization of additional money, or determination of alternative courses of action)? Or should we more logically provide food for thought that will establish a favorable climate for later follow-up?*

7. What material should be **withheld** from the briefing itself, but be available for reference during the question-and-answer period?

   *Are there some things that are not essential to the briefing objectives, but that you should have "in reserve" in the event they are raised by someone in your audience?*

for selection of subject matter

8. Finally, submit all resource material to the **"Why?"** test.
   *Back off and look at it objectively, as a disinterested person. Examine each item selected for inclusion in the briefing and ask yourself "Why is this to be used? What contribution will it make to the briefing's objectives?" Whatever cannot withstand this critical examination should be eliminated. This is a somewhat painful exercise because there is a natural tendency to include related material that is especially interesting or meaningful to the person preparing the briefing. However, in reality, this same information may hold very little interest for the audience, or it may dilute the ideas essential to the accomplishment of the briefing's objectives.*

Magic formula? No! Common sense? Yes! The logical and careful analysis of material to be selected for a briefing, always with the audience and the objectives in mind, is essential to the effective preparation of a briefing.

**PRACTICE EXERCISES**

1. Following the questions listed in this section, identify in writing all resource material you will include in the briefing you have selected. Make a supplementary list of material you may want to refer to during the question-and-answer period. Use the Resource Material Selection Worksheet, which follows, as an aid in this process.
2. Have your group of co-workers review this. You should be able to justify the inclusion of each item to their satisfaction as well as your own, or you should seriously consider eliminating it.

## RESOURCE MATERIAL SELECTION WORKSHEET*

Title or Subject: _____

Briefer(s): _____

Time and Place: _____

Date: _____

1. What is the *object or purpose* of the briefing?
   a. Parts of Preliminary Plan to consider _____

   b. Specific reference material needed _____

   _____

   _____

2. What should be *covered*? What can best be *eliminated*?
   a. Parts of Preliminary Plan to consider _____

   b. Specific reference material needed _____

   _____

   _____

3. What amount of *detail* is necessary?
   a. Parts of Preliminary Plan to consider _____

   b. Specific reference material needed _____

   _____

   _____

4. What *must* be said if the objectives are to be reached?
   a. Parts of Preliminary Plan to consider _____

_____

*For your convenience, a copy of *Resource Material Selection Worksheet* will be found at the back of this book.

## Resource Material Selection Worksheet (continued)

---

   b.  Specific reference material needed _____

_____

_____

5.  What is the *best* way to say it?
   a.  Parts of Preliminary Plan to consider _____

   b.  Specific reference material needed _____

_____

_____

6.  What kind of audience *action or response* is required if the objectives are to be met?
   a.  Parts of Preliminary Plan to consider _____

   b.  Specific reference material needed _____

_____

_____

7.  What material should be *withheld* from the briefing itself, but be available for reference during the question-and-answer period?
   a.  Parts of Preliminary Plan to consider _____

   b.  Specific reference material needed _____

_____

_____

8.  Finally, submit all resource material to the *"Why?"* test.

_____

BRIEFING PREPARATION STEPS

1. Establish objectives for the briefing.

2. Analyze your audience.

3. Prepare a preliminary plan for the briefing.

4. Select resource material for presentation.

5. **Organize material for effective presentation.**

6. Practice the presentation in advance.

Briefing Presentation: Step 5

## STEP 5: ORGANIZE MATERIAL FOR EFFECTIVE PRESENTATION

Once material is selected in line with the Preliminary Plan, it must be organized into an effective presentation which will fit your abilities, reflect your honest beliefs, meet your objectives, and satisfy the needs of your audience. Note that this is the first step where a specific presentation outline is discussed. All other steps have been preparatory, leading up to this one.

Although there are many different formats, most presentations can be broken down into three major parts: introduction, body, and conclusion. We have chosen to call these three parts: (1) State the Idea, (2) Develop the Idea and (3) Restate the Idea. Each serves a very specific purpose and requires a distinct approach.

### State the Idea (Introduction)

There are two primary purposes in the Introduction. The first and perhaps the more important purpose is to *sell the audience on listening to your briefing.* This is probably the most critical time in the briefing. With most audiences, there is a natural curiosity at the beginning of a presentation. Whether or not the audience retains that curiosity or interest will depend to a large degree upon your opening remarks and how effectively you convince them that it will be worth their while to listen attentively to what you have to say.

The second and more obvious purpose, of course, is to introduce the subject or purpose of your briefing, in other words, to *State the Idea.* It is vital to use simple and accurate words, and to make this part of your presentation interesting and *brief.*

### Six Sample Approaches

The specific approach you use will depend upon the subject matter, the presentation time allowed, the audience, and you yourself. Here are six of the many different ways in which a subject might be introduced. For illustration we will use the briefing on "Need for Increased Training in the Company," as identified in Preliminary Plan, Sample 4.

1. **Direct statement** of subject and why it is important to the audience.

    *Ladies and gentlemen, we are at a particularly critical time in the life of our company. In order to advance or even to survive in this highly competitive industry, we must substantially upgrade our human capabilities. There are two ways in which we can do this; we can hire the necessary talent from the labor market or we can train the talent we already have to meet the new challenges. Experience has shown that the first way is extremely expensive. Further-*

*more, testing has shown that such talent is extremely scarce in our area. The second alternative, a systematic, well-planned program of continuing training for our present personnel, appears to be the logical one in meeting this critical need. Let's examine the factors that lead us to this conclusion.*

2. **Indirect opening** on some *vital interest* of the audience and a statement connecting our objective with that vital interest.

   *How much time did you spend last week correcting or redoing the work of some of your subordinates because they lacked sufficient knowledge or understanding of what was expected—one hour, two hours, five hours, more? Could you have spent that time more productively working at your own level, on some new business, or even on the golf course or in a leisurely evening at home with the family? How much of your time and effort might have been saved if your subordinates were adequately trained? How much more time and effort will you have to expend in the future if steps are not taken today to meet the rapidly growing demand for increased technological and managerial knowledge and skill? How will an expanded company training program give you greater flexibility and allow you to use your time more efficiently? Let's see!*

3. **Vivid example** or comparison leading directly to the subject.

   *Last month our top management had to forgo bidding on a potential 10-million-dollar contract because analysis showed that under present operating conditions we do not have a sufficient number of trained personnel to handle the production nor can we hire or train them fast enough to meet the time requirements of this contract. How many such contracts can we afford to lose to our competitors? The time to act is now. This means instituting a far more comprehensive training program.*

4. **Strong quotation** relating to the subject, one that will be particularly meaningful to the audience and establish some rapport between you and them.

   *The editors of DUN'S REVIEW conducted a study several years ago of what 300 top corporate executives throughout the country generally agreed were the ten best-managed companies in U.S. industry. Their findings showed that these ten companies had six characteristics in common which contributed to their success. Without regard to priority, they are: (1) Abundant working capital, (2) A truly decentralized corporate structure, (3) An effective communications system, (4) High executive salaries and employee benefits, (5) Willingness to risk impressive sums of money on research and development, and (6) AN ACTIVE TRAINING PROGRAM THAT KEEPS NEW MANAGERS CONTINUALLY PRESSING TO THE FORE AND ESTABLISHED MANAGERS ON THEIR TOES.\**

   *Each one of these factors is worthy of in-depth analysis and discussion. For our purposes today, we'll concentrate on this last factor, the advantages of a first-rate training program.*

5. **Important statistics** having to do with the subject.

   *The cost of obsolescence and depreciation of facilities and equipment is a necessary expense we know we must plan on. Last year, in our company, depreciation amounted to some 10% of the value of our capital assets.*

   *In our business, trained personnel are also a major asset, and we must realize that a comparable allowance should be made if we are to maintain a labor force which is up-to-date, informed, and knowledgeable about all the advances being made in production. It is extremely important that we do this if our company is to grow and take on a large*

---

*From "What Makes a Best-Managed Company?" *DUN'S REVIEW AND MODERN INDUSTRY*, December 1963.

*share of the market. Yet our investment in terms of training hours per hours worked was only 0.4% last year.*

*While other factors are obviously related to human depreciation costs, these facts do raise some interesting questions about whether we are making a sufficient investment in training our people effectively to meet tomorrow's challenges. Let's examine this question in greater depth.*

6. **Story** illustrating the subject, provided it has a direct application and is not merely contrived for entertainment purposes.

   *A classic space-age cartoon shows a spaceship on its way to some distant planet with the question coming from somewhere inside, 'What dya mean, it won't flush!' Now this appears to indicate a lack of sufficient training on the part of someone—the astronauts, the assemblers, or the engineers. So that we won't get caught with OUR pants down, let's examine OUR training requirements.*

Any one of these introductory approaches might be appropriate under one set of circumstances and inappropriate under a different set of circumstances. Let's consider your approach as the presentor. Since the image you project is of vital importance in getting the attention of the audience, the method you choose of stating the idea should be the one you can do most effectively, if you can do this without violating the principles of audience reaction. Let's face it! Very few presentors can tell a story the way Bob Hope can. Therefore, the story or anecdote approach might be a poor choice for some. Normally, you would select just one approach for the first step, Stating the Idea, bearing in mind both requirements: selling the audience on listening to the briefing and introducing the subject.

## Develop the Idea (Body)

The idea, once stated, needs to be explained in whatever detail is necessary to accomplish the objectives of the briefing. Illustrations or actual performance of a new machine or procedure are more effective and usually much easier to understand than an explanation in words alone.

Meaningful illustrations play a critical part in the effective presentation of the body of the briefing. These may or may not be presented visually, but generally will carry more impact if they are. (The supplement to Chapter 3, "Selection and Use of Audio-Visual Aids," contains a listing of effective aids to understanding.) The proper use of illustrations is, in fact, so important that if a choice must be made, it is usually better to eliminate some of the specific facts than to cut out illustrations of the more important facts. The other facts can be added with supplementary handout material or at a subsequent briefing, or may be referred to during the question-and-answer period. Illustrations give impact and emphasis to the major points.

Once again, let me stress that it is essential to keep your particular audience in mind as you plan your methods of Developing the Idea. Their understanding of the subject, their particular interest in certain aspects of the subject, their individual and collective likes and dislikes of certain methods of presentation should guide you in selecting the means of interpretation. You may want to use:

a. *Examples* illustrating a projection of the idea in operation (flow charts, anecdotes, specific results, *pro forma* material).

b. *Reiteration* of the main idea in the same words or in different words to help summarize, to drive the idea home, and to ensure that the listeners will remember the point.

c. *Statistics,* if used sparingly and presented as simply as possible.

d. *Comparisons* with similar or dissimilar types of operations, ideas, and so forth.

e. *Testimony* of experts, witnesses to events, users of the product or procedure—such evidence should not be overused; the expert's reputation must be justified; and the credibility of the witness or user should be firmly established.

In this second step, Develop the Idea, you should consider also what would be the *minimum* amount of information necessary to get the *main ideas* across and what the best methods and most logical sequences might be. You should also decide whether to handle audience questions during the course of your presentation, at some specified point(s), or at the close of the briefing. Techniques for handling questions or discussion will be covered in Chapter 4, "Making the Presentation."

### Restate the Idea (Conclusion)

The conclusion, which is the weakest part of most briefings, deserves just as much or more planning than the rest of the presentation. It gives you the opportunity to sum up and stress the main ideas you want the audience to remember, to integrate and tie together various conclusions, and to suggest agreement and appropriate action.

```
   1. Summary of main ideas

 + 2. Appeal for action

 + 3. Review of purpose
      of presentation
 _____

   = GOOD CONCLUSION
```

Any idea you want the audience to remember needs to be repeated *from three to ten times* in a briefing, either verbatim or expressed in other words with a slightly different slant. Repetition, if handled intelligently, does not insult your audience, despite the common belief that it does. Obviously, you must exercise judgment in your use of repetition, but you are naive indeed if you expect the majority of the members of your audience to remember any point, even a significant one, if you make that point only once. Thus, to Restate the Idea is essential to the conclusion of the briefing.

The conclusion should not be lengthy, but it should be vivid and to the point in terms of what you want the audience to carry away with them. It is well to give the audience a strong indication of when you are starting the conclusion by using some such phrase as:

*Let's review the main points we've covered.*

*To sum up these factors, . . .*

*Our prime purpose today has been to . . .*

*Reflecting on what we have discussed, . . .*

Such a phrase, and you are certainly not limited to one of these, makes it quite evident that you are planning to wind up the briefing. This has a tendency to bring the audience back on target and gives you an opportunity to make certain they have gotten the main ideas and are prepared to respond or act in an appropriate manner to accomplish your objectives.

Each of the three steps to effective presentation organization has a definite purpose and needs specific attention. Failure to give each the attention it needs could result in the downfall of an otherwise well-prepared briefing. If you use all the imagination you have at your command, always keeping your specific audience in mind, you can come up with an effective method of presenting your material. Whatever format you choose, you will find it follows essentially the same three steps:

1. State the Idea   2. Develop the Idea   3. Restate the Idea
   (Introduction)         (Body)              (Conclusion)

A technique known as *Storyboarding* is a very useful device for organizing a presentation. (See Gus Matzorkis's article in Appendix A for an explanation of the approach.)

**PRACTICE EXERCISES**

1.  In outline form, organize your material for the briefing you are preparing, as illustrated in *Guidelines for Organizing Material for Presentation* and the *Presentation Worksheet* which follow. Prepare the outline in rough form so that it can be changed if necessary.

2.  Have your co-workers read your outline, tell you their reactions, and suggest ways for increasing its effectiveness. Carefully consider their ideas, but remember that it is *your* briefing, and you should use those items and approaches that will work best for you.

## GUIDELINES FOR ORGANIZING MATERIAL
## FOR PRESENTATION

---

**State the Idea** (Introduction)

Identify the plan for introducing the briefing. It should be short and have two primary purposes: (1) selling the audience on listening to the briefing and (2) introducing the subject or purpose of the briefing. Some approaches are:

1. DIRECT STATEMENT of subject and why it is important to the audience.

2. INDIRECT OPENING on some VITAL INTEREST of the audience.

3. Vivid EXAMPLE or comparison leading directly to the subject.

4. Strong QUOTATION relating to the subject.

5. Important STATISTICS having to do with the subject.

6. STORY illustrating the subject.

**Develop the Idea** (Body)

1. Following the MAIN IDEAS listed in the Preliminary Plan, identify your plans for supporting or interpreting these ideas. (See text for suggested methods of doing this.)

2. List and, if practical, diagram possible audio-visual aids to be included and indicate how and when they might be used. These must be only IN SUPPORT of briefing presentation.

3. Indicate how any questions and/or discussion are to be handled.

**Restate the Idea** (Conclusion)

Identify the plan for conclusion. It should:

1. Provide a summary of main ideas and objectives.

2. Have a possible direct appeal for action, belief, or understanding.

3. Review vividly the idea or purpose of the entire presentation.

# PRESENTATION WORKSHEET*

Title or Subject: _____

Briefer(s): _____

Date, Time and Place: _____

## General Considerations

1. How will the room be arranged (seating arrangements, lighting, namecards, etc.)? _____

   _____

   _____

2. How and when will the audience be notified of the briefing? Approximately how many will attend? _____

   _____

   _____

3. What equipment, aids, and supplies will be required? How will they be transported to the briefing location? _____

   _____

   _____

4. What handout materials will be required? What arrangements have to be made for them? How and when will they be distributed at the briefing? _____

   _____

   _____

5. How and when will you handle audience questions? _____

   _____

   _____

_____

*For your convenience, a copy of *Presentation Worksheet* will be found at the back of this book.

**Presentation Worksheet (continued)**

---

**Presentation Outline**

| Time | Content | Methods/aids/examples |
|------|---------|----------------------|
| | STATE THE IDEA (Introduction) (Sell the audience on listening; introduce the subject) DEVELOP THE IDEA (Body) RESTATE THE IDEA (Conclusion) | |

**Presentation Worksheet (continued)**

---

**Presentation Outline**

| Time | Content | Methods/aids/examples |
|------|---------|-----------------------|
|      |         |                       |

BRIEFING PREPARATION STEPS

1. Establish objectives for the briefing.

2. Analyze your audience.

3. Prepare a preliminary plan for the briefing.

4. Select resource material for presentation.

5. Organize material for effective presentation.

6. **Practice the presentation in advance.**

Briefing Presentation: Step 6

## STEP 6: PRACTICE THE PRESENTATION IN ADVANCE

The preparation of a briefing is not complete until you have actually rehearsed the presentation in a practice session. Most briefers have had the experience of outlining a good briefing on paper only to have it fall flat in actual presentation. Many factors are responsible for a failure of this kind. The following are among the more common ones:

1. The spoken words don't flow as smoothly as they seem to on paper.
2. The presentor loses continuity because of some distraction.
3. The mechanics of handling audio-visual aids interfere with the presentation flow.
4. The presentor is not as knowledgeable about the subject matter as presumed.
5. Someone in the audience asks a question the presentor did not anticipate or is not prepared to answer.

6. The audience is cold and unresponsive.
7. The place where the briefing is given does not lend itself to the type of presentation planned.

Rather significantly, all but the last two of the above factors are directly attributable to a lack of sufficient preparation or practice on the part of the presentor. Even the last two can be avoided or at least minimized if foresight is used.

Practice will not assure success, nor will it make a good briefing out of a poorly prepared one. However, it can and should:

1. Give you more self-confidence and poise, making the audience more willing to place credence in the subject matter.
2. Identify flaws or gaps in your material.
3. Provide familiarity with your material so that the right words come naturally and spontaneously.
4. Allow you to utilize the aids so that they will strengthen, not interfere with, the actual presentation.
5. Make it easier to anticipate potential questions, particularly ones that might prove troublesome.

## Methods of Practicing

There are three primary methods of practicing the briefing before actually presenting it. Any one or a combination of all three can be extremely valuable.

1. *Give the presentation aloud to yourself.* Retreat to a room alone with your notes and the audio-visual aids you will be using. Imagine your actual audience is there and make the presentation just as if they were. This will provide the feel of the material and the presentation flow, practice in the use of the aids, and an opportunity to identify elements of the material that may need polishing.

Practicing the Presentation

2. *Use a tape recorder.* The ready availability of inexpensive cassette tape recorders makes this a particularly useful tool. Even though it does not reproduce the full effect of the briefing, it does provide you with an opportunity to hear how you sound and see whether your ideas are coming through as you want them to. Usually you are your own best critic. This technique allows you to listen to the presentation more objectively, from the audience's viewpoint, and discover flaws that might otherwise escape you. If the presentation is critical enough and if the proper facilities are available, a video tape recording would be even more valuable, since you could watch as well as listen to your presentation. (See the article by Tom Sechrest in Appendix A for techniques in use of video recorders.)

3. *Give a dry run.* Have some knowledgeable co-workers, friends or, perhaps, even some representative members of your design audience sit in on your practice presentation. Although frequently more difficult for the briefer than the actual presentation, this kind of rehearsal is probably the most effective way to try out your techniques, make sure your ideas are getting across as you want them to, and learn how to field rough questions on the subject. It is much better to get egg on your face in this situation which is only a rehearsal than to do so with your regular audience. Be certain, however, that your dry-run audience gears itself to react in a manner typical of the individuals with the knowledge, interest, and attitudes of your design audience.

"Practice makes perfect" is a tired cliche and rather ridiculous in this context because it is unlikely that you will ever give a *perfect* briefing. However, "No practice means disaster" is a realistic statement. The best prepared briefing in the world can fail to achieve its objectives if it is not presented effectively. It is a rare individual indeed who can make a well-timed and forceful presentation without first practicing it. Chapter 4 will cover the techniques of effective presentation.

## PRACTICE EXERCISES

1. Practice your briefing, using as many of the suggested methods as possible.

2. Use the *Briefing Evaluation Guide* which follows. This is an effective tool to give you audience reaction in a dry-run practice session. (We have also printed the Briefing Evaluation Guide on perforated pages at the back of this book which you may remove and reproduce.)

3. Make adjustments in your final presentation based on the results of your practice sessions.

## BRIEFING EVALUATION GUIDE*

Presentor:_____ Evaluator:_____

## CONTENT

### Introduction

1. How good was it in arousing interest in the briefing?
   Outstanding_____Good_____Fair_____Weak_____

2. Was the purpose of the briefing made clear?
   Yes_____Somewhat_____No_____(?)[†]_____

Comments:

### Body

1. Did the main ideas come through clearly?
   Yes_____Somewhat_____No_____(?)_____

2. Were the supporting ideas and illustrations:
   Interesting?     Yes_____Somewhat_____No_____
   Varied?          Yes_____Somewhat_____No_____
   Directly related? Yes_____Somewhat_____No_____

---

*For your convenience, a copy of *Briefing Evaluation Guide* will be found at the back of this book.

†The items marked "?" are for those instances where the evaluator does not consider himself technically competent to judge and says, in effect, "I don't know."

**Briefing Evaluation Guide (continued)**

---

3. Was the presentation appropriate for the audience (as identified by briefer)?
   Yes_____Reasonably so_____No_____(?)_____

Comments:

**Conclusion**

1. Did it sum up main ideas and purposes?
   Yes____Somewhat____No_____(?)_____

2. How effective was it in encouraging action, belief, understanding?
   Outstanding_____Good_____Fair_____Weak_____

Comments:

**General**

1. How would you grade the briefing?
   Outstanding_____Good_____Fair_____Weak_____

2. Were the objectives of the briefing likely to be reached?
   Yes_____Probably_____No_____(?)_____

Comments:

**Briefing Evaluation Guide (continued)**

---

## PRESENTATION

### Audio-Visual Aids

1. Were they suited to the topic and the audience?
Yes_____Reasonably so_____No_____

2. Would they be visible to everyone and easy to follow?
Yes_____Reasonably so_____No_____

3. How effective was the use of these aids?
Outstanding_____Good_____Fair_____Weak_____

Comments:


### Platform Techniques

1. Poise: Was the presentor in control of the situation?
Yes_____Reasonably so_____No_____

2. Were posture and movements appropriate?
Yes_____Reasonably so_____No_____

3. Were gestures effective?
Good_____Fair_____Overdone_____Ineffective_____

4. Was relationship with the audience effective (eye contact, etc.)?
Outstanding_____Good_____Fair_____Weak_____

Comments:

**Briefing Evaluation Guide (continued)**

---

**Vocal Techniques** (Check more than one, if necessary)

1. How about pitch and quality?
   Good_____Too high_____Too low_____   .
   Monotonous_____Harsh_____Nasal_____

2. How about rate and intensity?
   Good_____Too fast_____Too slow_____Too loud_____
   Too soft_____Monotonous____

3. Did he or she speak clearly and distinctly?
   Yes_____Reasonably so_____No_____

Comments:


**General**

1. How did you feel about the speaker's overall presentation?
   Outstanding_____Good_____Fair_____Weak_____

2. Make any general comments you feel would be helpful.

# SELECTION AND USE
# OF AUDIO-VISUAL AIDS

We have examined the basic steps in preparation of an effective briefing. Careful attention to each of these steps is essential for a well-designed presentation. Now we will take a look at audio-visual aids, illustrative material that can add significantly to the impact of your briefing and the understanding your audience will gain.

## INTRODUCTION

Although most briefings can be made substantially more interesting and effective by incorporating audio-visual aids, probably no part of the presentation is misused more. However, careful thought, imagination, and professional assistance where needed can ensure their proper and effective use. The following basic principles and guildelines should be considered when you select the aids you will use.

### Audibility or Visibility

Can it be seen or heard satisfactorily by everyone? In other words, have you taken into consideration the size of the room, the number of people, the seating arrangements, lighting, and so forth? An aid that cannot be properly seen or heard is worse than no aid at all; it will irritate and distract your audience. A competent illustrator can assist you in determining the best type and size of aid to use in your particular situation.

If professional advice is not available, try out your aids in the actual or a similar location far enough in advance so that you can change them if necessary.

### Accessibility or Availability

Will the aid or equipment for showing the aid be available to you where you will be giving the briefing? Can it be made available at a cost (money, time, or convenience) that will not exceed its relative value to the briefing? Can it be stored in the order of use, but out of sight of the audience before and after it is used? A model of a particular piece of machinery can be an effective aid, for example, but it distracts the audience if it is left in view throughout the briefing. If such aids cannot be placed completely out of sight, you can keep them covered when they are not actually being used. The same principle, of course, applies to the use of charts. You can insert blank sheets of paper between them so that the charts will be covered when you want to turn the attention of your audience to something else.

### Adaptability

Does it fit? The aid should be selected to illustrate some particular point you are trying to make. Don't make your points fit the aid you have available (the tail should not wag the dog). Is the presentation of the aid smooth and well rehearsed? Any difficulties you have in using the aid will distract the audience and detract from the effectiveness of your message. If you are the type of person that machines hate (and there are many such people), avoid the use of mechanical aids.

### Appropriateness

It is appropriate to the audience? To the occasion? To the subject? To the speaker himself? Our example of the use of cartoon figures (see p. 19) illustrates one kind of inappropriateness

Determine the kind of reaction you want from your audience and the kind of image you want created of you and of your presentation.

## Arresting Quality

Will it gain attention and keep it on the subject matter rather than on the gimmick itself? A group of nude figures might be a novel way of drawing attention to a group of significant financial figures. But, which set of figures will your audience remember after the briefing? The aid should emphasize the *subject matter* you want remembered. It does not necessarily have to be flashy. The use of color (in moderation), heavy lettering, underlining, arrows, and other such methods can emphasize those factors you particularly wish to bring out.

## Auxiliary Nature

*The most important thing* to keep in mind is that an aid should be just what its name implies; it should *support* the briefing, not be the center of attention. A good briefing should be able to stand on its own *without any aids*. The aids should strengthen what you are saying, but should require your interpretation. Otherwise, you become merely a robot that operates the aids. This can raise a question in the subconscious minds of some members of the audience as to your competence as a briefer.

## TYPES OF AIDS AND EQUIPMENT: SELECTION, PREPARATION, AND USE

There are many different types of charts and many types of aids other than charts that can be used most effectively. Several of these are described as *Effective Aids to Understanding* in the supplement to this chapter. There are also examples of different kinds of charts, some good, some poor, for illustration purposes. In addition, you will find six rules for *Effective Use of Audio-Visual Aids*. We will discuss some of the more frequently used types of aids and equipment.

## Charts

Since charts represent the most frequently used briefing aids, we will deal with them first. Many of the points covered here will, of course, apply equally well to other types of aids. Let us examine first various methods of displaying charts and then certain rules concerning the selection and preparation of a briefing chart.

**Flipover charts** prepared on large sheets of paper and attached at the top to an easel by a clamp are the most commonly used kind. Each chart is flipped over when you finish discussing the material it displays. The size of the charts can vary considerably, depending upon the material to be displayed, the size of the group, and the location in which it is to be used. Normally, a skilled illustrator will be required to prepare the chart either by drawing it directly on the paper or by using printing or some other reproduction process. The cost of each such chart can run from as little as $3 to as much as $100, depending upon the amount of artwork and the process used. These charts can usually be rolled up and carried fairly easily.

**Chart cards** are similar to the flipover charts except that they are mounted on posterboard or heavy cardboard. They are sturdier, look neater and usually last longer than the flipovers. However, they are more costly and extremely clumsy if you have to carry them. Chart cards can be displayed on an easel, on a table leaning against something, on the tray of a chalkboard, or mounted on the wall. Sometimes they are covered with a sheet of plastic, both for protection and to permit writing over and above the original chart material. If the plastic cover is used, be careful where the chart is placed so that it will not reflect light into the eyes of your audience.

**Desk-top charts** are prepared on 8½" by 11" paper or card stock and are displayed in a binder similar to a three-ring notebook (actually, a notebook could be used if supported properly). These are used on top of a desk or table for a presentation to one person or to a group of not more than five.

**35mm slides,** or larger ones if desired, can be prepared by photographing charts and displaying them on a screen through a slide projector. This technique is particularly effective if your group is large and might have difficulty seeing other types of charts. It has some disadvantages; it usually requires a darkened room (which means subsequent loss of eye contact with your audience), limits the flexibility you have with a chart sequence, and may require coordination with a projectionist, although some projectors can be operated by the briefer with a remote control cord.

**Overhead transparencies** are becoming increasingly popular as a practical, inexpensive, and extremely versatile method of displaying charts. Display material is put on a specially treated sheet of plastic by a photographic process, through the use of an inexpensive copying machine (any 8½" by 11" black copy can be transferred this way) or by drawing directly on the plastic. With an overhead projector (there are several excellent ones on the market), material can be displayed on a screen or even on a light-colored wall. This kind of projector can be used with room lights on and can be operated by the briefer from the front of the room. This arrangement enables you to maintain eye contact with your audience. You can write on the transparencies much as you would on a chalkboard, construct a chart step by step through the use of overlays, and can turn the projector off when it is not in use. There are some problems with keystoning of the image, blocking of view by the projector or the presentor, and occasional difficulties in adjusting transparencies. These drawbacks are relatively minor, however, compared to the advantages. Any sales representative for one of the projectors will be glad to demonstrate its wide versatility.

**Ten Rules for Selection and Preparation of Charts.** Let us say at the outset that these are "80%" rules. We can all point to valid exceptions to any one of them. If you do choose to make an exception, however, be sure you have a sound reason for doing so.

1. The chart should be kept *simple in detail* and word usage.

2. For most situations, there should be *no more than ten lines* per chart and at least *two minutes* should be allowed for its use. Although numerous exceptions to this can be cited, the time control is still valid. It takes twenty to thirty seconds for your audience to focus on the content of a chart if it is in any way complex or a type of chart they don't ordinarily use. We confuse and frustrate the members of the audience if we flip the chart too fast. Also, if the point is important enough to be illustrated, it is important enough to justify spending sufficient time so that the audience can absorb the facts. Many briefings include too many charts. Timing will give you a standard by which the number of charts can be controlled. If you do make an exception to this rule, be certain that you *need* all the charts.

3. The chart should *clarify an idea better* than speech alone could.

4. The copy should present *highlights only*, requiring interpretation by the briefer. Don't give the whole story in the copy of the chart. Don't compromise the effectiveness of a chart in order to use it also as a handout. (See comments later about handouts.)

5. The chart should have *large, clear, bold, uncrowded letters and lines.*

6. It should be *large enough* for all to see easily, and stand high enough so that lettering at the bottom isn't blocked by the audience.

7. It should be an *accurate representation* of the facts depicted, not misleading in any way. If you are not a statistician, ask one to examine any figures or graphs to be sure they are truly representative and that no flaws in interpretation exist.

8. *Color* should be used *only to highlight important points.*

9. The chart should be *carefully made,* not thrown together. Not all charts need to be prepared professionally, but they should be neat and a creditable representation of you, your department, and the ideas you want to project.

10. It should be *sturdy and easily portable.*

### Handouts

Handouts in connection with a briefing can be an effective means of increasing learning and retention by your audience *if* careful thought has been given to their preparation and use. Otherwise, they are costly and wasteful, and may even detract from the objectives you are trying to acomplish. The following guidelines may help.

1. Normally, handouts should be made available *following* the presentation unless you intend to have the audience refer to specific portions of them during the briefing. Otherwise, you are inviting your audience to look at the handouts when they should be listening to you. As you go along, you can build an interest in the handouts which will be available after the session. You can indicate something new or different to be found in the handout without letting the audience satisfy their curiosity immediately.

2. When it is desirable (by design or by request) to reproduce 8½" by 11" copies of your briefing charts to be distributed

as handouts, make copies of *only* those charts that are *vital* for later reference. Reproducing all charts may be costly. Furthermore, the larger the package of reproduced charts is, the more likely it is to be relegated to the bottom drawer or the circular file. Conversely, the smaller the package and the more meaningful the items in it, the greater likelihood there is of its being used as a tool for future reference.

3. Reproduced briefing charts that are not completely understandable by themselves should be accompanied by interpretive remarks, printed perhaps on the reverse side. Remember, most charts should require additional interpretation by the briefer. Such interpretation may be lost after a time lapse. Reverse-side notes can help in recall.

4. Be certain that handouts will make a significant contribution to the accomplishment of your objectives; otherwise, *don't use them*.

## Chalkboard

The chalkboard is one of the most useful, most available and least expensive forms of visual aid equipment you can use in a briefing. It is not an easy aid to use, however, and requires advance planning and practice if it is to be used effectively.

*Advantages in the use of a chalkboard*

1. Flexibility: It offers plenty of space and can be changed relatively easily; words, diagrams, sketches—all can be used.

2. Feeling of spontaneity: The audience can get the feeling that this is the latest information, so current that a chart could not be prepared. If you become skilled in the use of a chalkboard, you can make modifications based on questions or comments or even on new ideas that may come to you or be suggested by audience reaction during the course of the presentation.

3. Progressive development: Starting with an initial step (as in a flow chart) or with a simple base (as in a model diagram),

an idea, process, or design can be developed progressively on the board as you explain each step.

4. Audience involvement: You can ask for inputs from the audience (particularly effective in developing a concept) and write them on the board. The very process of putting something on the board gets audience involvement. They try to think with you and anticipate what you are going to put there.

5. Change of pace: Writing on a chalkboard adds variety to the briefing, helps to maintain audience interest, and gives the audience a chance to make notes.

6. Nervous release for you. A straight lecture can build up tension in the presentor; writing on the board can help to relieve the pressure.

*Tips for the effective use of a chalkboard*

1. Write legibly and neatly—combine longhand and printing occasionally for emphasis. Write much larger than usual.

2. Hold chalk at approximately a 45° angle and apply sufficient pressure to make a fairly heavy line. If the chalk squeaks, break it in half. A metal chalk holder (available in stationery stores) will reduce or eliminate squeaking, breaking and dust.

3. Stand clear of chalkboard and remove all obstructions. Make sure all members of the audience can see the board.

4. Material that is prepared and put on the chalkboard ahead of time should be covered until you need it.

5. Allow enough time for the audience to copy or study the material developed on the chalkboard.

6. Maintain a flow of talk while you write on the chalkboard. Avoid as far as possible talking directly to the board. Always maintain audience contact.

7. Avoid writing lengthy material on the board during the session. Step back every few seconds when writing a series of items so that the audience can see what you're writing.

8. You can give an impression of spontaneity and have a good memory jogger, if you draw a faint outline of a diagram, model, or words in advance with pencil or charcoal. It will not be seen by the audience, but it will provide you with guidelines for the chalk.

9. For material you prepare in advance, use a straight edge or compass if possible. Here again, lightly penciled lines can help to provide a neat, uniform appearance.

10. Erase material on the board when it is no longer needed. "Always wash the breakfast dishes before serving lunch."

*Alternatives to a chalkboard*

1. Blank chartpad. In some ways this is even more practical than the chalkboard. You can use a felt pen or grease pencil to write on a pad of blank chart paper that is mounted on a sturdy easel. It has three distinct advantages over the chalkboard: (1) you don't need to erase; you just flip the page; (2) pages can be removed and mounted on the wall with masking tape for continued reference; (3) there is no chalk dust.

2. Blank transparencies or roller attachment for an overhead projector. Many such projectors come equipped with an attachment for a roll of blank plastic on which you can write with a grease pencil or fine point felt pen and then roll on when you have finished. If the attachment is not available, a supply of blank transparencies can serve the same function. Plastic page protectors (available in any stationery store) makes a creditable substitute if you do not have ready access to a supply of blank transparencies.

3. Blank card stock for desk top briefings.

4. Other types of display boards. There are many kinds of boards (flannel, hook tape, magnetic, etc.) that can be used to provide variety to your presentation. Audio visual dealer catalogs or exhibits should provide many new ideas if you feel the need to go beyond the standard equipment identified here.

## Visual Projection Equipment

There are so many different types of visual projection equipment, from the simple to the very sophisticated, on the market today that we won't attempt to describe them all. We will briefly identify those most commonly used by briefers.

We have already mentioned two types of readily available projection equipment, the *slide projector* and the *overhead projector*. In addition, the *filmstrip projector* is similar to the slide projector, with a series of frames on a continuous strip of film. It can be equally as effective except for the fact that you cannot change the sequence of pictures so easily. A fourth type, the *opaque projector*, is used only when it is necessary to project something from a book or some other form from which it would be impossible or impractical to make either a slide or a transparency. Usually the opaque projector must be operated from the back of the room, and it requires a darkened room. A fifth type of equipment, the *motion picture projector,* can be very effective in conveying an idea, particularly when the briefer does a good job in preparing the audience by an introduction and in conducting a meaningful discussion or summary after its showing.

Companies producing this equipment publish considerable literature describing it and its use. Ask your local sales representative or write to the company for more information on it.

*General rules when using visual projection equipment*
1. Set up equipment before the session. Minimize the mechanics of operation you will need during the session.
2. Check the seating arrangements and remove any visual obstructions.
3. Designate an assistant to help in light control, if necessary.
4. Leave equipment intact (don't try to rewind film or rearrange slides) until after the audience is dismissed.

*Hints on the use of a motion picture projector*
1. Have spare lamps and fuses available.

2. Check for proper threading, focus, and position before the session begins.
3. Allow time for warming up the projector's amplifier.
4. Adjust sound to a conversational level—increase the volume after you start, if necessary.
5. Adjust the film to start at the beginning of the acknowledgments (title and credits).
6. Use the following procedure in starting a film showing:

   - Turn on the machine for warmup.
   - Turn on the projection lamp.
   - Darken the room.
   - Make minor sound and focus adjustments, if needed.
   - Start showing film.
   - Turn off the sound as soon as THE END is shown.
   - Turn on the room lights.
   - Turn off the projection light.
   - Start the discussion immediately.

*Hints on the use of overhead and slide projectors*

1. Have a spare lamp available.
2. It may be desirable to lower some of the houselights, depending upon how powerful your projector is.
3. Position the first frame or slide to be used and check for proper focus.
4. Place all frames or slides in order, handy to the projection area, before the start of the session.
5. Coordinate actual projection onto screen with related presentation discussion. If there is an interval where slides do not need to accompany the discussion:

   a. Overhead Projector—Turn off the light until the next frame is needed.
   b. Slide Projector—Cut out, before the presentation, a piece of cardboard the size of the slide and put it in the

slide tray between the frames to make a break. This will provide a temporary screen blackout and make it unnecessary to turn the machine off.

c. Be sure to turn the houselights on if the blackout period will be lengthy.

6. Turn on the room lights when the final frame or slide is shown. Then turn off the projector, but leave the fan on to cool the bulb.

## Recorders

Occasionally the use of some sort of recording can be an effective way of presenting material. Recordings of key people who cannot be present may add prestige as well as content value and a change of pace in the briefing. Sometimes the recording of an interview or a conference can be used to advantage. The sound (or a combination of sight and sound) of machinery in operation, a test being conducted, or some other sound effect may add dramatic impact and ensure meeting your objectives.

**Record players** may be useful if you have professional recordings, but they are not very practical if what you need to present is your own material.

**Audio tape recorders,** for most purposes, are the easiest, least expensive, and most practical type of recorder. In addition to the larger, more standard models, there are many small, inexpensive recorders of high quality that can be operated by either electricity or batteries. The audio cassette recorder is a particularly useful device which has the added advantage of being virtually "operator proof." These recorders are adequate to meet most briefing requirements. The audio tape recorder is particularly useful with a slide presentation. It can be timed to a running commentary or to provide appropriate sound for what is being seen on the screen.

**Video tape recorders** are somewhat more ambitious aids than most briefings warrant. However, the tremendous technological advances that have been made in recent years in the use of this medium warrant special attention to its use. (See the article by Tom Sechrest in Appendix A.)

## Multimedia

The mixed use of several different media, such as overhead transparencies, flipchart, recorder, handouts, worksheets, has the advantage of involving many senses of the audience. This adds impact as it sustains a higher level of interest and increases the likelihood of retention. You are limited here only by the extent of your imagination and appropriateness to the situation. Two cautions: (1) don't overdo it to the extent that your media distract from the message; (2) allow substantial time for practice to ensure that the mechanics of handling several media don't interfere with the message either.

## Storyboarding

This is a particularly effective technique for organizing a presentation and coordinating the use of supporting audio visual aids. (See Gus Matzorkis's article in Appendix A.)

We have covered only the more practical types and uses of the audio-visual aids available. Your ingenuity, knowledge of your audience, careful preparation, and practiced use of suitable aids are what will support an effective presentation. However, remember that as their name suggests they are *aids to,* not replacements for, a well-prepared and well-presented briefing.

### PRACTICE EXERCISES

1. Outline as many different ways as you can to illustrate each key point you wish to make in the briefing you are preparing. (For ideas, see *Effective Aids to Understanding* and the sample charts included at the end of this chapter, and keep in mind the suggestions made throughout.) Discuss these possibilities with your fellow-workers and consider any additional suggestions they may have.

2. Select the aids most suitable. If you plan to present charts or other illustrations, make a rough sketch of how the finished aid should look. Check it out for effectiveness with your fellow-workers.

3. If you plan to make the finished illustrations yourself, go over your sketches with an experienced illustrator and use any suggestions offered. If you are having the work done, allow sufficient time so that you can check the artist's rough draft and make any corrections or suggestions before the final product is started. Doing so will save time and money and avoid unnecessary trouble.

## SUPPLEMENT TO CHAPTER 3

### EFFECTIVE AIDS TO UNDERSTANDING*

The following aids can be used to make the subject easier to understand or more interesting, and to promote the kind of thinking that will help you accomplish your objectives:

### 1. Charts

To direct thinking; clarify a specific point; summarize; show trends, relationships, and comparisons.

Information charts or tabulations should usually be prepared in advance to ensure that all points are covered and covered accurately.

### Types of charts

**Highlights**, straight copy, emphasizing key points.

**Time-Sequence** (historical), showing relationships over a period of time. May be in seconds or centuries. Can use pictures or graphs.

**Organizational,** indicating relationships between individuals, departments, sections, or jobs.

**Cause-and-Effect,** e.g., picture of bottle plus driver plus auto equals wrecked auto.

**Flow Chart,** to show relations of parts to finished whole or to the direction of movement. A PERT (Program Evaluation and Review Technique) chart is a flow chart.

**Inventory,** showing picture of object and identifying parts off to the side by arrows.

---

*For your convenience, a copy of *Effective Aids to Understanding* will be found at the back of this book.

**Dissection,** enlarged, transparent, or cut-away views of object.

**Diagrammatic or Schematic,** reducing complex natural objects by means of symbols to simple portrayal, e.g., radio wiring diagram.

**Multi-Bar Graph,** using horizontal or vertical bars representing comparable items.

**Divided-Bar Graph,** a single bar divided into parts by lines to show the relation of parts to the whole.

**Line Graph,** using a horizontal scale (abscissa) and vertical scale (ordinate), e.g., showing number of overtime hours being worked each month.

**Divided Circle,** pie graph, used in the same way as the divided bar.

**Pictograph,** a pictorial symbol representing comparable quantities of a given item, e.g., stacks of coins representing comparable costs of different phases of an operation.

## 2. Illustrations, Diagrams, and Maps

To clarify a point, emphasize trends, get attention, or show relationships or differences.

## 3. Motion Pictures

To show motion, give overall view or impression, or show actual operation.

## 4. Film Strips or Slides

Same as for motion pictures. They are especially helpful when motion is not required or desirable. They are not expensive and can be studied as still pictures.

## 5. Samples or Specimens

To show the real object.

## 6. Models

*Small-scale*—To permit showing an operation without using large quantities of material, to make a large operation visible, or to show a project to be completed.

*Large-scale*—To make an object large enough to permit handling, identify small parts, or see internal operation.

## 7. Exhibits

To show finished products, the results of good and poor practices, attract attention, arouse and hold interest, and adequately display one idea. (Use life, motion, color, or light to help attract attention.)

## 8. Worksheets

To provide "hands on" experience by audience in performing certain actions; to provide a carry over to the job.

## 9. Manuals, Pamphlets, Instruction Sheets, Circular Letters, Outlines and Bulletins

For standard information and guides, for reference and background.

## 10. Cartoons, Posters, Signs

To attract attention and arouse interest.

## 11. Photographs, Textbook or Magazine Illustrations

To tie discussion to actual situations and people, provide current interest, or show local activities.

## 12. Case Studies

To bring together for specific situations, the principles, practices, and procedures which are being explained, interpreted, or formulated by the group. It is much easier to visualized a procedure if you "Take the case of Mr..."

## 13. Examples and Stories

To break monotony or tension, fix an idea, get attention, illustrate or emphasize a point, clarify a situation, or break away from a delicate or ticklish subject.

## 14. Demonstration

To show how to carry out a suggested method or procedure.

## 15. Field Trips

To present subject in its natural setting, stimulate interest, blend theory with practicality, and provide additional material for study.

## RULES FOR EFFECTIVE USE OF AUDIO-VISUAL AIDS

---

**Rule One:**    NEVER USE A VISUAL AID BEFORE AN AUDIENCE UNTIL YOU HAVE REHEARSED WITH IT.

- Make sure it works properly.
- Arrange parts in proper sequence.
- Practice handling the aid.

**Rule Two:**    MAKE CERTAIN THE AID IS A HELP RATHER THAN A HINDRANCE TO COMMUNICATION.

- It should be simple, and clear, and it should demonstrate a single idea.
- Use contrast, color, or other ways to emphasize or clarify the main points.
- Handle the aid only when you are making a direct reference to it.

**Rule Three:**    KEEP TALKING WHEN APPROPRIATE.

- Do not interrupt your talk when you change slides or handle aids.
- Use the aid to support what you are saying. Avoid talking to support the aid.

**Rules for Use of Audio-Visual Aids (continued)**

---

**Rule Four:** SPEAK WITH MORE VOLUME THAN IS NOR-
MALLY REQUIRED—PROJECT!

- Remember that the listener's attention is divided.
- In a darkened room, more volume is required to hold attention.

**Rule Five:** DON'T STAND BETWEEN YOUR LISTENERS AND THE VISUAL AID.

- Stand to one side.
- Use a pointer.
- Face and talk to the audience.

**Rule Six:** USE VISUAL AIDS IN PROPER SEQUENCE.

- Don't show them until you are ready to use them.
- Don't pass samples around. Show the samples to the group as a whole or show them after the presentation.

SAMPLE CHARTS

HIGHLIGHTS CHART

Poor copy
Too much boldface
Not enough white space
Monotonous

PRINCIPAL CONTI

ASSOCIATE CONTI

PARTICIPATING

CONTRACTORS

GUIDANCE &

NAVIGATION

DESIGN APPROACH

SPECIFIC GUIDE LINES

HIGHLIGHTS CHART
Good Copy

• NO DIRECT LABOR CHARGE PRORATION

• STANDARD INDIRECT LABOR PRORATION AGREEMENT

• JOB CLASSIFICATION

• CUSTOMER APPROVAL FOR MAJOR CHANGES

RELIABILITY MILESTONES & STATUS

SEQUENCE CHART

Poor Copy
Too much detail
Print too small
Difficult to trace time flow

GO-AHEAD
SUBCONTRACTOR-SUPPLIER RELIABILITY ADMINISTRATION INITIATED
PRELIMINARY MISSION PHASE RELIABILITY APPORTIONMENTS COMPLETE
PRELIMINARY SUBSYSTEM RELIABILITY APPORTIONMENTS COMPLETE
RELIABILITY PROGRAM PLAN SUBMITTED
RELIABILITY PROGRAM PLAN REVIEWED
RELIABILITY/CREW SAFETY DESIGN CRITERIA DOCUMENTED
EDUCATION COURSE-COMPUTER METHODS OF DESIGN ANALYSIS IMPLEMENTED
QUALIFICATION-RELIABILITY TEST PLAN SUBMITTED
QUALIFICATION-RELIABILITY TEST PLAN REVIEWED
FIRST RELIABILITY PROGRAM REVIEW
REVISED RELIABILITY PROGRAM PLAN SUBMITTED
FIRST QUARTERLY PROGRESS REPORT SUBMITTED
REVISED QUALIFICATION-RELIABILITY TEST PLAN SUBMITTED
FIRST QUALIFICATION STATUS LIST SUBMITTED
FIRST MONTHLY FAILURE SUMMARY SUBMITTED
SUPPLIER SYMPOSIUM ON HI-REL PARTS CONDUCTED
SECOND REVISION TO QUALIFICATION-RELIABILITY TEST PLAN
EDUCATION COURSE-FUNDAMENTALS OF RELIABILITY MATHEMATICS IMPLEMENTED
SECOND QUARTERLY PROGRESS REPORT SUBMITTED

| JAN | FEB | MAR | APR | MAY | JUN | JUL | AUG | SEP | OCT | NOV | DEC | JAN | FEB | MAR |
|-----|-----|-----|-----|-----|-----|-----|-----|-----|-----|-----|-----|-----|-----|-----|
| | | | | | | 1962 | | | | | | | 1963 | |

# TIME PHASING

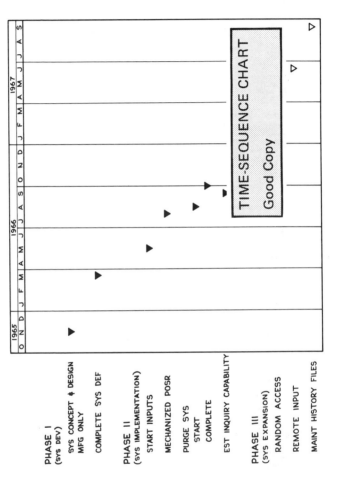

| | 1965 | | 1966 | | | | | | | | | | | 1967 | | | | | |
|---|---|---|---|---|---|---|---|---|---|---|---|---|---|---|---|---|---|---|---|
| | O N D | J | F M A M J J | A | S O N D | J | F M A M J J | A S |

PHASE I
(SYS DEV)
SYS CONCEPT & DESIGN
MFG ONLY
COMPLETE SYS DEF

PHASE II
(SYS IMPLEMENTATION)
START INPUTS
MECHANIZED POSR
PURGE SYS
START
COMPLETE
EST INQUIRY CAPABILITY

PHASE III
(SYS EXPANSION)
RANDOM ACCESS
REMOTE INPUT
MAINT HISTORY FILES

TIME-SEQUENCE CHART
Good Copy

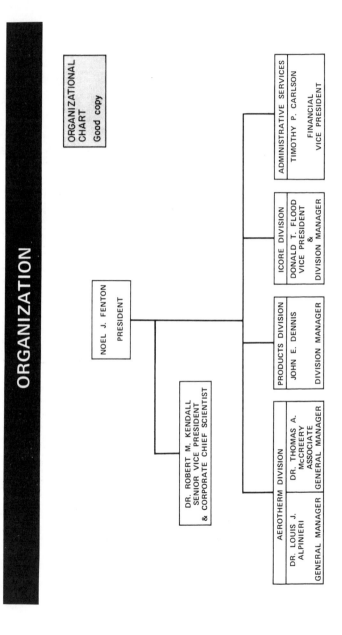

ORGANIZATION

ORGANIZATIONAL CHART
Good copy

NOEL J. FENTON
PRESIDENT

DR. ROBERT M. KENDALL
SENIOR VICE PRESIDENT
& CORPORATE CHIEF SCIENTIST

AEROTHERM DIVISION

DR. LOUIS J. ALPINIERI
GENERAL MANAGER

DR. THOMAS A. McCREERY
ASSOCIATE GENERAL MANAGER

PRODUCTS DIVISION

JOHN E. DENNIS
DIVISION MANAGER

ICORE DIVISION

DONALD T. FLOOD
VICE PRESIDENT
&
DIVISION MANAGER

ADMINISTRATIVE SERVICES

TIMOTHY P. CARLSON
FINANCIAL
VICE PRESIDENT

Courtesy of ACUREX Corporation

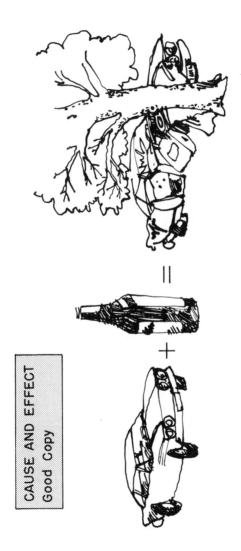

CAUSE AND EFFECT
Good Copy

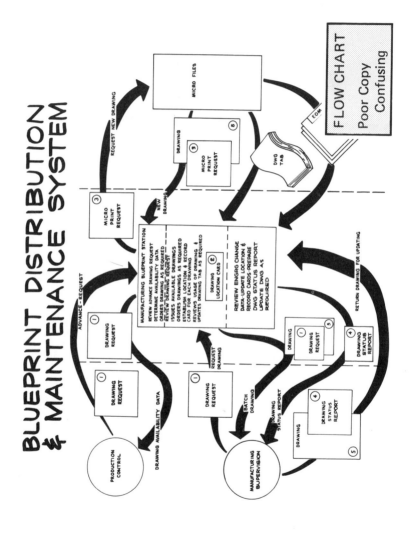

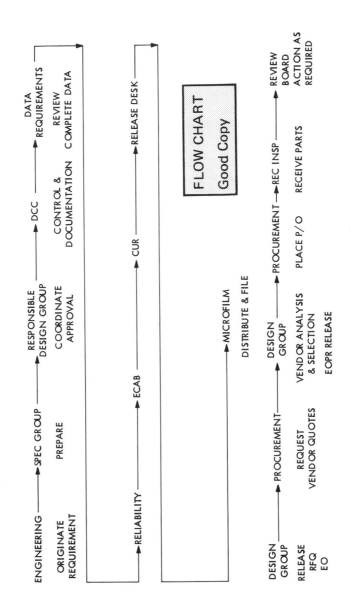

ENGINEERING DATA INTERFACE ACTIVITY
PROCUREMENT SPEC FLOW

FLOW CHART
Good Copy

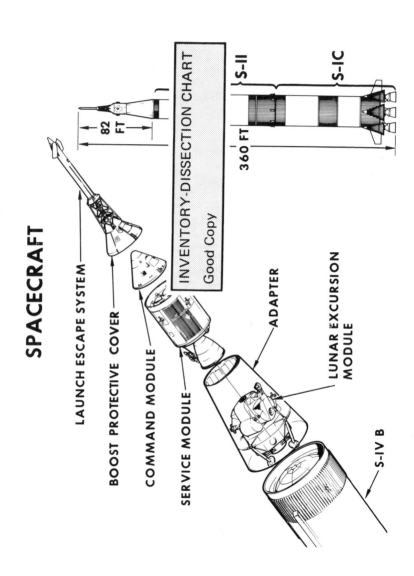

SPACECRAFT

LAUNCH ESCAPE SYSTEM

BOOST PROTECTIVE COVER

COMMAND MODULE

SERVICE MODULE

ADAPTER

LUNAR EXCURSION MODULE

S-IV B

82 FT

360 FT

S-II

S-IC

INVENTORY-DISSECTION CHART

Good Copy

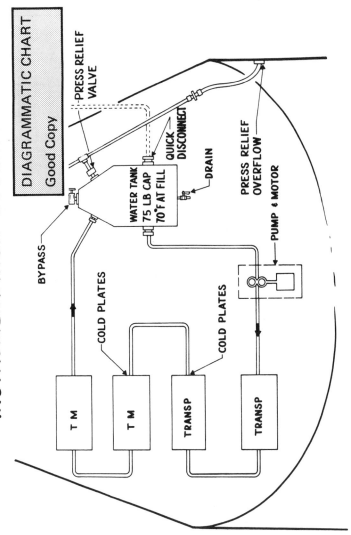

INSTRUMENTATION COOLING

DIAGRAMMATIC CHART
Good Copy

PRESS RELIEF VALVE

QUICK DISCONNECT

DRAIN

PRESS RELIEF OVERFLOW

WATER TANK
75 LB CAP
70°F AT FILL

PUMP & MOTOR

BYPASS

COLD PLATES

COLD PLATES

T M

T M

TRANSP

TRANSP

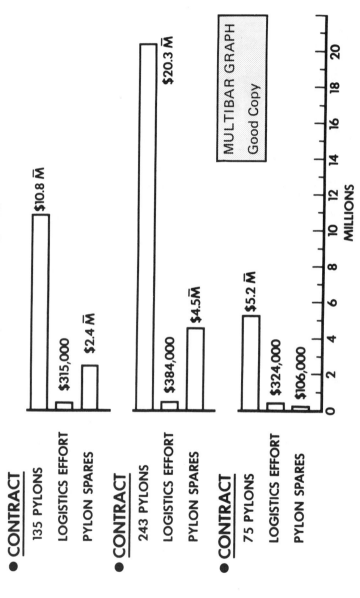

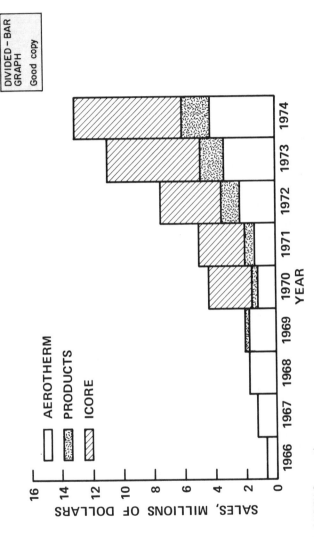

SALES HISTORY

DIVIDED – BAR
GRAPH
Good copy

AEROTHERM
PRODUCTS
ICORE

SALES, MILLIONS OF DOLLARS

16
14
12
10
8
6
4
2
0

1966  1967  1968  1969  1970  1971  1972  1973  1974

YEAR

Courtesy of ACUREX Corporation

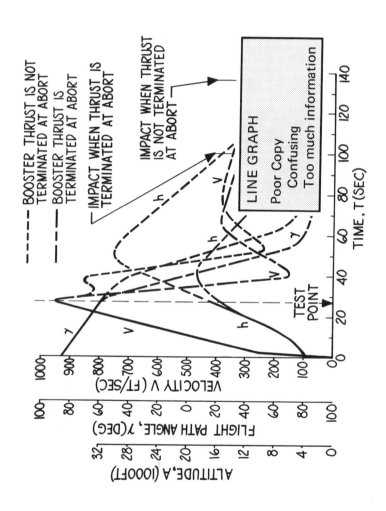

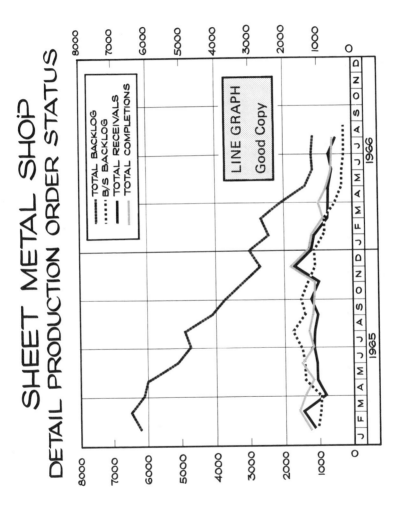

SHEET METAL SHOP
DETAIL PRODUCTION ORDER STATUS

LINE GRAPH
Good Copy

TOTAL BACKLOG
B/S BACKLOG
TOTAL RECEIVALS
TOTAL COMPLETIONS

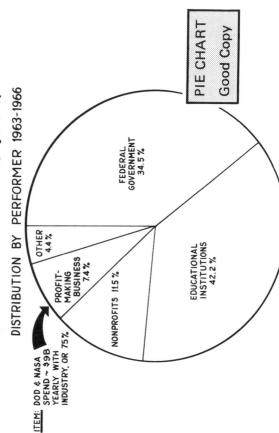

CUMULATIVE TOTAL RESEARCH & DEVELOPMENT OBLIGATIONS OF NONDEFENSE/NONSPACE GOVERNMENT AGENCIES (OGA)

DISTRIBUTION BY PERFORMER 1963-1966

PIE CHART
Good Copy

FEDERAL GOVERNMENT 34.5%

EDUCATIONAL INSTITUTIONS 42.2%

NONPROFITS 11.5%

PROFIT-MAKING BUSINESS 7.4%

OTHER 4.4%

ITEM: DOD & NASA SPEND ~ $9B YEARLY WITH INDUSTRY, OR 75%

BASED ON NOMINAL $2 BILLION ANNUAL TOTAL R&D MARKET

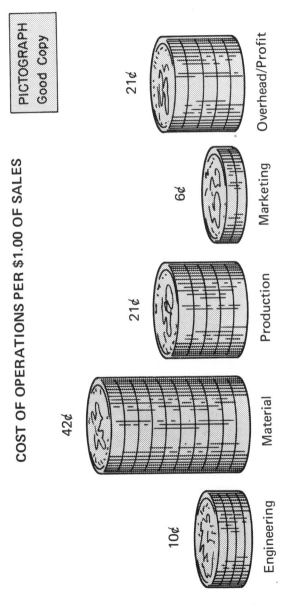

COST OF OPERATIONS PER $1.00 OF SALES

PICTOGRAPH
Good Copy

42¢

Material

10¢

Engineering

21¢

Production

6¢

Marketing

21¢

Overhead/Profit

# 4

# MAKING THE PRESENTATION

Now that we have thoroughly studied the preparation of a briefing and have examined the various approaches to selection and use of audio-visual aids, we come to the actual presentation. This is the moment of truth when you will find out how successful you are and whether your work will be justified. However, even though the briefing itself has been carefully prepared, there still are several additional reponsibilities you must assume if you want the best possible reception for your presentation.

## PRELIMINARY ARRANGEMENTS

"It's my job to give the briefing, not play janitor and convener. I've got enough to worry about without that. Let someone else take care of the details." This sounds like a reasonable attitude except for one thing. You have a vested interest in seeing that the details are handled properly. If the "someone else" who has the responsibility fails to follow through effectively, it is *your* briefing that will suffer. You should give some personal attention to each of the following to make certain the job has been done properly. (See *Preliminary Arrangements Checklist* which follows for guidance.)

### Room Arrangements

There are many ways to arrange the seating in a room depending on the size and shape of the facility, the size and nature of the audience, the type and method of briefing, and the kind of participation you want from the members of the audience. Here are some of the more conventional seating arrangements together with some factors to be considered about each. The recommended square footage to be allowed for each participant is based on a rectangular room without visual obstructions. Additional allowances should be made for unusual shapes, columns, or other things that might interfere with vision, hearing, or comfort. Also, in calculating room capacity, be sure to allow 40–100 square feet for the briefer and the required briefing accessories.

*Auditorium Style* (8 sq. ft. per person). Useful for large groups with a relatively short time period (± one hour) and where little or no writing or use of reference materials is required by the audience. Audience participation is normally limited to question-and-answer periods.

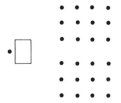

*Classroom Style* (16 sq. ft. per person). Useful for relatively formal situations where participants will need to write or make active use of reference materials. Audience participation is normally limited to question-and-answer periods.

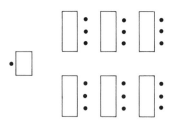

*Informal Style* (16 sq. ft. to 20 sq. ft. per person). Useful for small group (12 people or less) where a high degree of discussion is desirable.

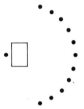

*U-Shape* (20 sq. ft. per person). Useful where individual eye contact with audience members, writing or use of materials by participants, and open, relatively informal, discussion are all desirable.

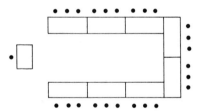

*Buzz Style* (20 sq. ft. to 24 sq. ft. per person). Useful when it is desirable to have small group discussions on all or parts of what is being presented. Round tables.

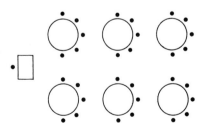

*Herringbone Style* (20 sq. ft. per person). More formal than buzz style; more informal than classroom style. Rectangular tables.

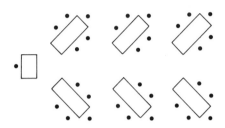

In addition to seating arrangements, there are several other concerns you should have about the meeting room. Are there enough chairs and tables? Is there adequate lighting without glare for you, your visual aids, and your audience? Is there sufficient ventilation, particularly if you anticipate several smokers? What is the potential for distractions in the room, outside of windows or doors, in adjoining rooms? What can you do to eliminate or compensate for them? Are there enough ash trays, pencils, scratch paper, and so forth? Would a name-card for each member of the audience, enabling you to address each by name, add to the effectiveness of your overall communication? Would it be desirable or practical to have coffee, soft drinks, or water available for the audience when they arrive or during the session? You can check to make sure that the details are taken care of before your arrival.

### Equipment and Aids Arrangements

Are all required equipment and aids in the room and in the proper position for use when needed? Are they working properly? Is there a spare projection bulb available in case one blows? Are there electrical outlets in the room, and will they accomodate the type of power plug you have? (An oversight on this last relatively simple matter need happen only once to convince you of how important good preparation is.) Will an extension cord be available, if one is needed? Is a table for your projection equipment, display materials, etc., available and accessible for effective usage without significantly blocking the view of any of your audience?

## Handout Materials

Have they been well prepared and assembled in the proper order, out of sight, in sufficient number for your audience, and are they ready to be distributed at the proper time (not before)?

## Audience Notification

Has the audience been notified of the briefing? Have all the right people, those most concerned or those who can make the necessary decisions, been invited? What was the nature of the notification—letter, memo, phone call, meeting announcement, word-of-mouth? How much information did the notification give about the topic, its purpose, your qualifications, time of the meeting and how long it will take, etc.? How persuasive was the notification?

Out-of-Plant
Briefing

## Out-of-Plant or Out-of-Town Briefings

Who is coordinating local arrangements? Have you provided clear instructions as to your needs? Have arrangements been made to procure, ship, or carry with you all necessary equipment, charts, handouts, and so forth? Have you allowed sufficient margin on travel time to permit last-minute adjustments on meeting arrangements, if necessary?

You will be well repaid for the time and effort you spend on these preliminary arrangements. Many briefings have fallen apart because of failure to check out one of these relatively small details. The *Preliminary Arrangements Checklist* which follows should help keep you out of trouble.

## PRELIMINARY ARRANGEMENTS CHECKLIST*

Session:_____Briefer:_____ Coordinator:_____

Date & Time:_____Room:_____ No. Expected:_____

**Seating arrangement** (No. per table_____)

Auditorium_____Classroom_____Informal_____U-Shape_____

Buzz_____    Herringbone_____Other_____

**Room considerations**

|  | Satisfactory | Not Needed | Needs Attention | Comments |
|---|---|---|---|---|
| Chairs |  |  |  |  |
| Tables |  |  |  |  |
| Lighting |  |  |  |  |
| Ventilation |  |  |  |  |
| Distractions |  |  |  |  |
| Ash trays |  |  |  |  |
| Pencils/scratch paper |  |  |  |  |
| Name-cards |  |  |  |  |
| Coffee/soft drinks/water |  |  |  |  |
|  |  |  |  |  |
|  |  |  |  |  |
|  |  |  |  |  |

*For your convenience, a copy of *Preliminary Arrangements Checklist* will be found at the back of this book.

## Preliminary Arrangements Checklist (continued)

---

### Equipment / aids / handouts

| (List) | Satis-factory | Not Needed | Needs Attention | Comments |
|---|---|---|---|---|
| | | | | |
| | | | | |
| | | | | |
| | | | | |
| Electrical accessories (bulb, cord, plug, extension) | | | | |
| | | | | |
| | | | | |
| | | | | |
| | | | | |
| Supplies (chalk, eraser, felt pens, grease pencil, tape) | | | | |
| | | | | |
| | | | | |
| | | | | |
| **Audience notification** | | | | |
| | | | | |
| | | | | |

## PLATFORM TECHNIQUES

Platform techniques are designed to increase your ability to get your message across. Although some of our previous discussion has dealt with effective platform techniques, let's enumerate some specific guidelines in greater detail.

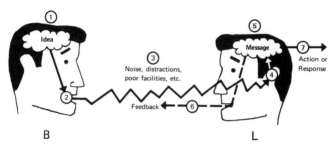

### Relationship to Audience

Communication is a two-way process. (See the accompanying diagram.) Our briefer (B) does not communicate merely by putting ideas (1) into words (2), no matter how accurate or well chosen those words may be. The listener (L) must be receptive to these words despite outside interference (3) so that what the briefer says not only reaches the ear (4), but enters the brain (5) in a form that is similar to the original idea (1). To test this interaction and understanding, the briefer (B) must get some form of feedback (6). Furthermore, *effective* communication must lead to some form of action or response (7).

Feedback, both positive and negative, can be received in many ways: by questions and comments from the listeners, by observation of nodding heads, facial expressions, vacant stares, attention or the lack of it, and so forth. Inherent in all of these is a sensitivity in the briefer, an awareness of the listeners' reactions.

A vital part of good communication lies in effective *eye contact*. For the inexperienced presenter, maintaining eye contact with the listeners can be difficult and frightening. It is much more comfortable to pick out a nice inanimate spot on the back wall and speak to it. However, the briefer who does this is losing out on one of the most valuable resources available for determining whether or not the message is going across— visual feedback. Also, confidence in you is much stronger if you look your listener in the eye when you are making your presentation. A personal relationship can be established with *each listener* so that the individual feels you are talking directly to him or her during part of the presentation. In addition, looking at a member of the audience (without staring) can be a very effective way of holding or regaining that person's attention if it has wandered.

Keeping Eye Contact

The most important component in **Comm⎯nication** is the **you** and, unfortunately, as in the spelling of the word here, it often is overlooked or missing. The entire presentation must be prepared and delivered in a manner that is understandable, interesting, and meaningful *from the point of view of the listener.* Isn't it amazing how stupid and short-sighted many audiences are for not seeing the situation in the same way we do? However, that's just the way most of them are and we must learn to deal with them on that basis. Come to think of it, if they weren't like that, there probably would be very little need for the briefing and possibly, very little need for our services. What a disturbing thought that is!

### Presentation Tools

Many presentation techniques have been identified. Two additional tools are worth consideration.

**Lectern or speaker's stand.** Depending upon the type and size of lectern, it can serve several purposes in addition to the obvious one of being a place for your notes:

1. Out-of-sight storage space for your aids and handouts with convenient access when they are needed;
2. Resting place for your hands, although you should avoid gripping it tensely;
3. Device for establishing a particular type of relationship with the audience. This is one of its most subtle, yet most effective, uses. Remaining behind the lectern continuously tends to establish a somewhat formal relationship, which is desirable at times. Moving to the side or in front of the lectern, in addition to providing a change of pace, tends to remove both the obvious physical barrier and an unseen barrier, making for a closer, more informal relationship with members of the audience. Moving back behind the lectern is a good way of focusing attention on the summary as a more formal part of the briefing.

**The pointer.** A pointer is a valuable aid in drawing attention to specific items on a chart. However, it is all too frequently a distracting toy. Beware of being a dueler, a pendulum, a tapper, or one of their many cousins. *Put it down when you're not using it!*

## Use of Body

Although a well-organized and illustrated briefing is the prime consideration, the manner in which you handle yourself during the presentation has a significant effect upon the audience. The effective use of your body frequently makes the difference between an apathetic reception and enthusiastic acceptance by your audience. The following suggestions should be kept in mind:

**Poise.** The poised speaker is the one who appears self-confident, relaxed, and capable of doing whatever the situation may call for. The experienced briefer can give this impression, even when feeling insecure, by paying attention to a few rather important details.

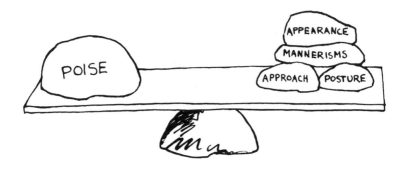

1. Dress, of course, should be in good taste, clean, and comfortable. If the occasion calls for informal dress, then use it. In general, it is not good practice to remove your jacket, loosen your tie, or roll up your sleeves when the relationship with your audience is reasonably formal. Keep in mind the kind of an image you want your audience to have of you and what you represent.

2. Approach to the speaker's position in the room should be deliberate and unhurried. Pause for a few moments after getting into position, smile comfortably, and look over your audience briefly before beginning to speak. This gives your audience a chance to focus on you and to feel that they don't frighten you (even when they do). Also, a moment of silence normally will do far more to attract the audience's attention than launching on your talk immediately.

3. "What do I do with my hands?" is a frequent question of the less experienced briefer. Very simply, you should do what feels most comfortable or seems most natural for you. Hold them loosely at your sides; raise one to your waist; put one hand in your pocket, both if you can do it without looking sloppy; fold them loosely in front of you; put them behind your back; rest them lightly on the lectern; hold

your 3 x 5 card notes in them; whatever seems best for you. But don't keep them in motion! They should be relaxed and should not draw attention. Be careful to avoid jingling coins or keys in your pockets.

4. Posture, whether standing, sitting, or walking, should be relaxed without being sloppy, and dignified without being stiff, reflecting the type of impression you want to give.

**Movements.** Full body movement periodically during a briefing serves a variety of purposes. Pacing back and forth is certainly no asset. However, deliberate, well-timed bodily movement can:

1. Relieve tension within you,
2. Draw attention away from a visual aid and back to you,
3. Break the hypnotic effect a stationary body has on the audience,
4. Change the mood or the pace of the presentation.

**Gestures.** Although they do not come naturally to most briefers, well-selected and well-timed hand gestures can add effectively to the type of audience reaction you want.

1. Some types of gestures:

   *Sweeping hand* illustrates covering a broad field; takes in entire audience, and so forth.

   *Vertical or chopping motion* emphasizes precise points; breaks an idea into parts. (Avoid waggling your index finger unless you intend to scold your audience.)

   *Palms out* says "Stop!"; rejects an idea.

   *Palms up* invites acceptance, open-mindedness, or participation.

   *Upturned fist* can draw audience to you; give aggressive emphasis.

2. Principles for the effective use of gestures:

   They should draw attention to the idea, not to the gesture itself.

Vary the types of gestures; don't over-use one or it loses its effect.

Proper timing of the gesture with the word or phrase it supports is vital.

They must be appropriate to the impression you want. The wrong gesture is worse than none at all.

Too many gestures limits their value. Very few people do overgesture, but control yourself if you have this tendency.

3. Experimenting with gestures:

Practice *in private* with different ones while talking, until they become comfortable. They may feel awkward at first, but they will become more natural with practice. Do not use them in a formal briefing (dry runs are OK) until they feel reasonably comfortable to you.

**Facial expression.** The face, in particular, should reflect the mood you want to create in your audience. The deadpan briefer will inspire neither interest nor enthusiasm. In almost any presentation, the following facial expressions will be appropriate at one time or another: serious, smiling, laughing, inquiring, doubtful. They should be lively, varied, and appropriate to the total situation.

**Distracting mannerisms.** Many people manifest distracting mannerisms of which they are totally unaware. The lip-licker, nose-patter, ear-tugger, scratcher, eyebrow-flutterer and head-bobber are only a few. A video tape recorder as a practice medium will reveal whether or not you belong in any of these categories. A sympathetic, honest co-worker can tell you whether you have any of these habits. Becoming aware of the problem is half the battle. It is up to you to overcome it. Many an audience's attention has been lost through fascination, amusement, or revulsion over a speaker's mannerisms.

**Be natural!** Don't try to present a foreign image to your audience. Your use of body must be comfortable and appropriate to you, to your audience, and to your subject.

## VOCAL TECHNIQUES

We have all tried to listen to a speaker who was presenting interesting material, who used good platform techniques, but who either irritated us or put us to sleep with an unpleasant or monotonous voice. Worse than this is the speaker you either cannot hear or cannot understand. You feel that your time is being wasted, and that whatever he or she is mumbling about probably isn't worth much anyway.

Very few people are endowed with the voice and oratorical ability of a Franklin D. Roosevelt, a Billy Graham, or a Charlton Heston. However, most speakers can increase their effectiveness substantially with whatever vocal tools they have if they give careful attention to certain basic elements of speaking and practice, using if possible a tape recorder or other means of playback.

1. **Pitch** or inflection refers to the tone of the voice. A conversational tone that is neither too high nor too low should be developed. It should seem natural for you and should be varied to prevent monotony. A voice pitch different from your normal speaking voice usually betrays nervousness and is distracting to an audience, particularly if they are familiar with the way you speak conversationally.

2. **Voice quality** problems, sounds that are nasal, thin, harsh, pinched, or breathy, may be difficult to overcome. However, if you are faced with any of these problems, you can minimize their effect. Speech instructors can suggest exercises to help overcome most of them.

3. **Intensity** is the force or loudness with which you project. Depending upon the size of your audience and the room arrangement, you should usually speak slightly louder than you would in normal conversation. The volume must be loud enough so that everyone can hear you, but not loud enough to overpower them. Variations in intensity add

dynamics. A soft voice, at times, can command more attention than a loud one. When you use it, however, speak somewhat more slowly.

4. **Rate** or tempo of speech is another important factor both in being understood and in making an effective presentation. The machine gunner usually loses an audience almost immediately, because it is impossible for them to catch everything being said. The foot dragger loses audience members nearly as quickly by boring them or irritating them with the way every phrase is dragged out. A tape recorder can be very helpful in showing you how good your timing is. This vocal problem requires a certain amount of discipline. *Variations* in rate can add considerably to the effectiveness of a presentation provided they are consistent with the mood you are trying to convey.

5. **The pause** is closely related to rate and can be effective in drawing attention to points that you consider particularly important. It should be used deliberately, however, so that you do not give the impression that you are groping for words.

6. **The "uh"** has long been the nemesis of public speakers. It results most often when thought processes interfere with speech processes. We fail to turn off the voice while we are thinking of what to say next. In our opinion, however, the "uh" is vastly over-rated as a speech problem. The occasional "uh" in speaking is not nearly so disastrous as some public speaking practitioners would have us believe. Speakers who have had this idea hammered into them often become "unglued" at the utterance of a single such sound. The continuous use of "uh" can be extremely distracting, of course, and anyone who has this problem should work at controlling it. The tape recorder, again, can be a valuable tool in helping to overcome it. A technique that has proved helpful in licking this problem is to *overdo* it. Every time as you practice and catch yourself saying

"uh", say it two or three additional times. Overemphasis on any fault draws it more clearly to your attention. Awareness is the most important step in learning to control it. Complete familiarity with what you are going to say should help to solve this problem.

7. **Voice drop** at the end of a sentence is another common fault among speakers. Without realizing it, many people let the last few words trail off to the point that they become difficult or impossible to hear. Therefore, the meaning of the thoughts they are trying to convey is frequently either lost or distorted. Since most offenders of this type are completely unaware that they do this, they can overcome this fault if it is called to their attention or if they recognize it in the playback of a tape recorder.

8. **Faulty pronunciation** is distracting to the audience and undermines their confidence in the speaker. When the audience has to take time out to determine what *has been* said, they are not giving their undivided attention to what *is being* said. If you are not absolutely sure of the proper way to pronounce a word, look it up in a dictionary.

9. **Enunciation** is the way a speaker articulates the words. "Mumbles" may have been an interesting comic strip character, but he or she is frustrating to anyone who really wants to know what is being said. Proper enunciation results in word projection that is clear, precise, and easy to listen to. The *consonants* (particularly the final ones) and not the vowels are the real key to effective enunciation. Here again, however, keep your audience in mind and speak much as they do: avoid sounding affected.

10. **Correcting faults** in use of the voice should be approached systematically. First, become aware of them through verbal critique or playback on a tape recorder. Study how to correct them, and seek help from others if necessary. Then, practice both incorrectly and correctly to get the feel of

how both methods sound and the difference between them.

How much attention you should give to each of these elements in the use of the voice only you can determine. Almost everyone can improve his or her presentation through careful attention to the elements discussed here.

## AUDIENCE QUESTION TECHNIQUES

Many an otherwise well-presented briefing has left a poor impression because of the speaker's inept handling of audience questions. Most industrial, business, and governmental briefings do provide for questions from the audience, and the way they are handled often has a greater effect upon the objectives of your briefing than does the presentation itself. You should, therefore, plan exactly how and when questions will be handled. Furthermore, you should anticipate the types of questions and types of questioners you may encounter, and plan how you will handle them.

### Audience Retention Curve

Before we consider question techniques, let us examine for a moment the audience-retention curve shown in the accompanying figure.

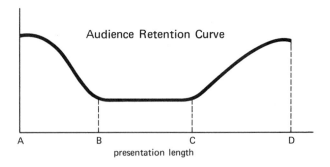

We are considering audience *retention* here, not *attention*. The two may be closely related, but are not necessarily identical. This curve shows which parts of the briefing are remembered best after the presentation is over. What is remembered, after all, determines whether or not you achieve the objectives of the briefing. While many unidentified factors, such as time, physical arrangements, predispostion of the audience, do affect this curve, we can make some reasonable assumptions.

Curiosity will probably make the introduction (A) gain a reasonably high degree of both attention and retention. Retention of a bad introduction, however, will be equally as high as that of a good one, and will color the audience's attitude toward the entire presentation. This factor makes it doubly important to do an effective job of gaining audience support during the introduction.

Following the introduction, a reasonably sharp drop in retention (B) is to be expected, even with a good presentation. Unless the briefing presentation content is very short and simple, a drop will occur, of that we can be sure. The bottom of the curve is not as smooth as the illustration would indicate, of course. There are peaks and valleys during the body of the presentation. In fact, a deliberate change of pace, a story, illustration, or audience activity, at this stage can help counteract both attention and retention drop. However, overall retention is almost certain to be lower during this portion of the briefing. For this reason, you must repeat important points during the body of the presentation.

The rise at the end of the curve (C) depends on the effectiveness of the briefer. The presentor who runs out of something to say and stops may have the retention curve stay at point C, at a low level. With a presentor who runs out of something to say and *doesn't* stop but rambles on with unnecessary repetition, the curve will drop off to nothing or, worse, may go up, indicating that the audience will go away remembering the dull ending and thinking that the entire briefing was a waste of time. However, an upward curve of rentention is highly probable if the audience can anticipate that the briefing is coming to a close. Thus the use of summary phrases suggested in Preparation Step 5 (see page 56) helps bring the curve up at this stage. With some sort of an indicator, those who have done some mental wandering will usually make a conscious effort to get back on the track and carry something of value away from the briefing (D).

What does the Audience Retention Curve tell us?

1. Audience retention is *lowest* during the body of the presentation, that portion of the briefing which normally takes the *largest* amount of preparation effort. This does not mean that we should fail to give it sufficient attention. It does mean, though, that this is not the place to introduce a key point and then drop it, if we want to have a high degree of later retention by the audience.

2. The introduction and, particularly, the conclusion require as much careful preparation as the body, possibly even more. Since these portions of the presentation are more likely to be retained, the primary points we want the audience to remember must be brought out here in an effective manner.

3. Attention should be given to techniques that can be used during the body of the presentation to stimulate the audience and keep attention and retention high. Changes of pace and a wise selection and use of audio-visual aids can help.

What does the Audience Retention Curve have to do with audience questions? It has a direct bearing on *when questions should be encouraged*.

1. Handling questions *during the briefing* can be the most effective method. This is the time when the questions asked are most meaningful to the questioner. Furthermore, the audience reaction gives you excellent feedback as to whether or not your message is being received correctly. In addition, questions along the way, which necessarily mean

active participation by members of the audience, will raise the retention curve. However, there are problems. If you are on a tight time schedule, lengthy questions may prevent your completing the briefing as you had planned. Unless the questions are of interest to all the audience, some of your listeners will feel their time is being wasted. A premature question may upset the way you had planned to develop the material. It takes a great deal of skill to handle questions effectively and still maintain both control and continuity of the briefing.

2.  Usually questions *follow the presentation* and, from the standpoint of accomplishing the briefing's objectives, this is often the *poorest* time for them. By turning this prime time over to the audience, you run the risk of having them leave the briefing remembering your difficulty in handling an embarrassing question or a curve someone may have thrown—someone who does not agree with your point of view. Even if there is no opposition, there is always the possibility that someone will bring up a point that might be very interesting but one that is totally irrelevant to the purpose of your briefing.

Sometimes established protocol prescribes that a question-and-answer period follow the formal briefing. If that is so, consider having the question period *before* you present the summary. Since the final minutes of a briefing are the prime time, why turn it over to someone else? A simple statement, "I'll be happy to answer any questions you may have, but I would like to hold the final two or three minutes for a summary" can give you the control you need. This will give you an opportunity to recover from irrelevant or embarrassing questions and send the audience away with *your* ideas, not someone else's.

As an interest-building technique, consider leaving out some important information from the formal part of the briefing, anticipating a question from the audience on it. If the audience does call this to your attention, they will get a feeling of more active participation. At the same time you can improve your own image by showing how effectively you can answer the question. Also, you will have some reserve ammunition to counteract any opposition point of view. If the question is not raised, the information purposely left out can be included in the summary.

## How to Conduct a Question-and-Answer Period

Since the manner in which you conduct yourself during a question-and-answer period can be one of the most critical factors in the success or failure of your briefing, here are some specific pointers you should keep in mind as you prepare for it.

Your attitude is definitely the most important single consideration. If you approach audience questions with the idea that someone is trying to put you on the spot or catch you in a mistake, you are bound to be defensive. If, on the other hand, you approach them with the idea that the audience is paying you a compliment by asking questions, implying that they are genuinely interested in gaining information that only you can provide, the difference in climate can be tremendous. In fact, this attitude can be completely disarming to anyone who might be trying to shoot you down. Actually, though, most people are sincerely interested and should have their questions answered in a manner that reflects a positive attitude.

It is often necessary to clarify what information the questioner desires. Without embarrassing the individual, you can repeat or rephrase the question to be sure of its true meaning. If it is a buckshot question (several questions in one), it is better to zero in on one precise point rather than to deal with the total question in generalities. For example, in the briefing on "Need for Increased Training in the Company" (Preliminary Plan, Sample 4), the following exchange might take place between a questioner (Q) and the briefer (B):

Q: *How much time should an employee spend in formal training?*

B: *Do you mean time spent during regular working hours?*

Q: Yes.

B: *That would vary considerably depending on the position and the circumstances. Could you give me an example of a case you might have in mind?*

Q: *Let's say a first-line supervisor of a group of auditors.*

B: *Usually a first-line supervisor will require about twice as much formal training as the average individual employee, since he or she needs both technical and managerial training. Therefore, an investment of 40 to 80 hours a year, depending upon specific requirements, would not be out of line.*

Evaluate the question in relation to the objectives of the briefing. You may have to pause for a few moments to think about it before attempting to answer. This is one of the real tests of a skilled briefer. The ability to make a quick mental evaluation and then proceed with an answer in line with that evaluation is a skill that requires continued disciplined practice. Controlling the urge to blurt out an immediate answer prior to such evaluation is a discipline that will provide a great advantage to the briefer. For example, in the Training briefing, a questioner might say, "You mentioned the ABC Company's training experience. Can you tell us how they got started on their program?" Even if the briefer is thoroughly familiar with the background in the other company, a detailed answer will do little to accomplish the briefing's objectives. Therefore, through effective mental evaluation, the briefer should comment very briefly on only those factors that have a direct bearing on the present company requirements. If the questioner is not satisfied, the briefer should offer to discuss it in greater detail after the briefing.

Learn to spot beforehand those potentially weak areas in the presentation so that you are not unduly embarrassed if they are challenged. A government official who was requesting an increased budget to provide for 800 additional employees did not help his cause when he was unable to give a satisfactory explanation of his department's inability to fill 200 vacancies already in existence. Careful, objective analysis on your part should reveal most of these areas of weakness. Listen objectively to your practice recording specifically to spot places where you may be challenged. Ask someone to play Devil's Advocate at a dry run. Then, either find satisfactory answers to questions that are difficult or, at the very least, determine what response you will give if such questions are raised.

Always give some sort of an answer, even if it is "I don't know." Each question asked must be dealt with in some way or your image will suffer in the collective minds of the audience. If the question asked will be covered later in the briefing, say so and give a brief answer, indicating that more detail will be given. Don't postpone the answer completely unless it will seriously impair the flow of the material. In fact, coming back to it later will reinforce the idea you had planned to get across. If the question is diversionary, leading the discussion away from the immediate course of your briefing, summarize briefly where the presentation was before the question in order to bring the audience back on target and ensure the continuity of the presentation.

## Difficult Types of Questions and Questioners

Whenever you open a question-and-answer period, you run a risk. You must develop skills for dealing with difficult types of questions and questioners. You should remember, however, that you have two primary responsibilities: first, to do justice to the material you are presenting, and second, to meet the needs of the *entire* audience, not of a single member of it unless, of course, that single member is a *key* individual who must make the decisions. Here are a few suggestions for dealing with some of the most frequent problems that arise.

**The argumentative individual** who attempts to get you into a one-to-one argument over a particular item or point of view must be turned aside. Keep this in mind: Even if you can nail that individual to the wall, you nearly always *lose* in such an argument. First, the arguer usually won't let go even when nailed. Second, an extended argument usually is of little interest to the rest of the audience, and you can lose them. Third, if you make that person look foolish, the rest of the audience is likely to identify with the individual and resent it.

Although there are exceptions, a person who argues in public is primarily seeking recognition, both from the briefer and from the rest of the audience. He or she sees it as an opportunity to demonstrate personal knowledge and capability or to air a particular gripe. What's the best way to deal with an arguer? *If recognition is what is really being sought, give it and get off it.*

"You raise some very interesting ideas, Tom. I'd like to take the time to explore it in more detail with you. Can we get together right after the meeting?" You might lose a few points but the outcome won't be nearly as disastrous as it would be if you were to stand still and trade punches.

If you do have a satisfactory answer for the question that won't antagonize your audience, try something like "Thanks for raising that question, Tom. I appreciate your point of view. We think that _____ will take care of that problem. I'll be glad to meet with you after the meeting and discuss it in more detail if you wish. Meanwhile, back at the ranch..."

**The curve or loaded question** is the one specifically designed to embarrass you or put you on the spot. It frequently is a question that cannot be answered, or it hints that you are trying to hide something. A typical curve might be "How do we know that the employee's performance would not have improved anyway, even without this training?

Like the arguer, this questioner usually is seeking recognition. He or she wants to play "Can you top this?" The question is asked, knowing what your answer will be, and with a counter-answer ready to fire when you are through. In addition to the techniques recommended for dealing with the arguer, try "turn-about" with this individual. "That's a very interesting question, Tom. What do you think about it?" If he still wants to play one-upmanship, you then have the home-team advantage. As with the argumentative individual, though, you have little to gain and much to lose by prolonging this discussion. Therefore, the more quickly you can get off the subject, the better.

**The windbag or long-winded questioner** rambles all over the lot or has to tell you his or her life's history before getting to the point. If allowed to drone on indefinitely, valuable time is lost, as well as the interest and attention of the rest of your audience. If possible, you should keep the individual from losing face when you intervene. This sometimes can be done by putting the issue off until after the briefing, as with the first two types of questioners. If this is not possible, try one of these methods:

1. If you can anticipate the question, jump in at your first opportunity and both ask and answer the question for the individual.

2. Pick up a word or idea he or she is expressing and show its relationship to something you or someone else has said previously.

3. As a last resort *only*, cut the questioner off in the interest of time. Let him or her down easily and give recognition if you can in order to avoid antagonizing either the individual or the rest of the audience. "That's very interesting, Jane. I wish we had the time to go into it as thoroughly as we should."

**If you don't have a good answer to the question**, *admit it* and either refer to someone who can provide it or offer to get it later. If you try to bluff, you not only do not satisfy the questioner, you raise doubts in the minds of the audience about the truthfulness of the entire briefing. We should admit it when we don't have the answer. Under the pressure of the moment, however, most of us are strongly tempted to protect our own egos by going ahead and hoping no one will catch our weakness. Resist that temptation!

**If you need time to think**, either to evaluate the question or because the question catches you a bit off guard (you can give an answer but need a few moments to collect your thoughts), pause for a moment. Instead of starting to answer immediately, hoping that your thoughts will catch up with your words, try one of the following gambits:

1. "Would you mind repeating the question (or expanding on it) so I can be sure I understand you?"
2. "That's a very good question. How do you feel about it?"
3. "That's an excellent question. Let's think about it for a few moments." (Pause until ready to answer.)
4. "That's an interesting question. How do some of the rest of you feel about it?"
5. If the question lends itself to it, write it or the answer on the chalkboard; this can give you thinking time.

Much more could be written on this most critical and controversial part of briefings. However, the principles and techniques suggested here can go a long way toward significantly increasing the success of your briefing presentations.

## SUMMARY

Effective presentation techniques will not make a good briefing out of poor material. The effort must first be on a thorough job of preparation so that the message itself is worthwhile. The more practice you can get in presentation techniques, particularly before a sympathetic yet critical audience, the more effective your presentation will be. Once you have a soundly prepared message, careful attention to the points covered in this chapter will make your presentation more meaningful and should make it much more likely that you will accomplish your briefing objectives.

# CONCLUSION

The concept of *Management by Objectives and Results* (MOR) is just as applicable to a technical presentation or briefing as it is to anything else we might manage. Whatever approach is followed must be focused on the *results* we want the presentation to accomplish. This text has established a basic approach to the briefing process as illustrated in the diagram presented initially in Chapter 1. Let's look at it again.

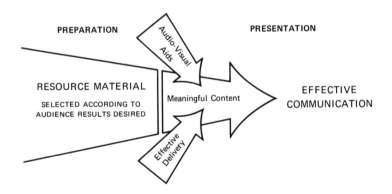

The subjects indicated by the three arrows leading to *Effective Communication* have been the subjects of the three principal parts of the text.

At the heart of the process is *Meaningful Content* which was developed in Chapter 2, "Steps in Preparation of a Briefing." They are:

1. *Establish the objectives* for your briefing. Determine *why* you are giving the briefing and *what you hope to accomplish* with it.

2. *Analyze your audience* in terms of their knowledge, attitudes, and ability to act. Keep this information always in mind as you prepare your briefing.

3. *Prepare a Preliminary Plan,* concentrating on the main ideas or concepts the audience must get if your objectives are to be met.

4. *Select the resource material* for your presentation, following your Preliminary Plan. Be prepared to justify the inclusion of each item in terms of your objectives.

5. *Organize your material* in a logical and effective manner for your audience, putting particular effort into a strong introduction and conclusion.

6. *Practice your briefing* before you meet your audience. Get the bugs out of the presentation so that you can present it as smoothly as possible.

In support of Meaningful Content are the *Audio-Visual Aids* which were discussed at length in Chapter 3. Careful selection, design, and use of such aids can mean change of pace, clarification, and stimulus in one of the most critical parts of a briefing.

*Effective Delivery,* explored in Chapter 4, serves to point up the importance of your *preliminary arrangements, platform techniques, vocal techniques,* and especially the *audience-question techniques.* We strongly recommend that you seek out opportunities to practice speaking in front of others. Join Toastmasters; take a public speaking course; volunteer for briefing assignments even if the thought of it paralyzes you. Experience is the only way to overcome or at lease reduce the size of those butterflies.

Finally, *EFFECTIVE COMMUNICATION* means getting the message across in a manner that will accomplish the objectives of the briefing. If your audience does what you want it to do as a result of your briefing, you have achieved effective communication—even if you have violated every principle covered in this text!

HERE'S TO BRIEFER, BETTER ORGANIZED, AND MORE EFFECTIVE RESULTS-ORIENTED PRESENTATIONS IN *YOUR* ORGANIZATION!

# BIBLIOGRAPHY

There are hundreds of books and other products on the market related to preparing and making presentations. I won't attempt to prepare an extensive list here, but I would encourage the interested reader to browse the libraries and management journals for additional ideas. I would like to identify a few, however, that I have found particularly helpful.

1. First, I want to offer a belated acknowledgement to the two resource books that were of most help to me in developing the original training program in 1962 which led, eventually, to the publication of the first edition of this book in 1968. Both published by McGraw-Hill Book Company, Inc., they are:

   Loney, Glenn M., *Briefing and Conference Techniques* (1959)

   Sandford, William P., and Yeager, Willard H., *Effective Business Speech* (1960)

2. Any book by Bob Mager is a worthwhile learning experience in addition to being a pure delight to read. All are published by Lear Siegler, Inc./Fearon Publishers, Belmont, California. The two that are most relevant to effective presentations are:

   Mager, Robert F., *Preparing Instructional Objectives* (1962), and *Measuring Instructional Intent* (1973)

3. A small but powerful book that was of particular value to me in working with highly educated engineers and scientists is:

   Weiss, Harold, and McGrath, J.B.N., *Technically Speaking: Oral Communication for Engineers, Scientists and Technical Personnel* (McGraw-Hill, 1963)

4. Gerry Nierenberg and Hank Calero have the best book out that I have seen on how to visually evaluate your listeners and assess your own body impact. It is published popularly as:

   Nierenberg, Gerard I., and Calero, Henry H., *How to Read a Person Like a Book* (Hawthorn Books, 1971; Pocket Book edition, 1973)

5. Addison-Wesley has three other books that I see as particularly useful to the individual in the "real world" who has to devote a significant amount of effort to formal communication as a part of the job. They are:

   Breth, Robert D., *Dynamic Management Communications* (1969)

   Broadwell, Martin M., *The Supervisor as an Instructor* (1970)

   Hays, Robert, *Practically Speaking—in Business, Industry, and Government* (1969)

## OTHER MEDIA

1. The film, "Anatomy of a Presentation" produced by Roundtable Films is the best one I have seen on the subject. It shows a novice presenter and what he has to go through to get ready for a presentation in a manner that most viewers can identify with. It then shows his actual presentation with an opportunity to evaluate it.

2. Addison-Wesley has three packaged training programs that are useful in this subject area. They are:

*Effective Presentations Instruction Kit* (available in direct support of *this* book)

*The Effective Speaking Filmstrip Series* (related to the Hays book, identified above)

*The Supervisor as a Classroom Instructor* (related to the Broadwell book identified above)

# APPENDIX A

# MANAGEMENT BY OBJECTIVES AND RESULTS
## A Snapshot

by George L. Morrisey

Management by Objectives and Results (MOR) is a common sense, systematic approach to getting things done that is based on principles and techniques that many good managers have been practicing for decades. In spite of the new jargon that has come into vogue, there is nothing mysterious about it. It does not require a manager to stop what he or she has been doing successfully for years and learn a whole new approach. That would be idiotic. It does require the manager to focus on *results* rather than activities, building on the strengths that he or she has developed over the years with modifications and additions as good judgment dictates. Its effect should be in reducing the "fire fighting" syndrome (it can never be eliminated) with greater attention being given to "fire prevention" as the mark of a truly professional manager.

The MOR process is simple—deceptively simple. Most managers and students of management do not have great difficulty in comprehending it intellectually. After all, it is perfectly logical. The difficulty comes in application because it does require systematic planning, an uncomfortable activity for many of us. It can be illustrated graphically as a horizontal funnel. As a process, it moves from the general to the specific. Its purpose is to subdivide an effort that is large and complex until it reaches a unit size that is manageable. Then it is integrated through a human process that promotes understanding, involvement, and commitment. (While a modest departure from the flow diagram that appears in my book, *Management by Objectives and Results*, this "funnel" is conceptually consistent with it.)

COMMUNICATION

1. *Roles and Missions* describe the nature and scope of the work to be performed. They establish the reason for the organization's existence. The "organization" can be the entire enterprise or the specific organizational unit(s) for which this particular manager is accountable. A description of the economic, functional, and other commitments involved, plus a determination of the philosophical basis for conducting the organization's affairs, are an integral part of this step in the MOR process. Once established, it is not likely to change unless there is a significant change in what the organization will be doing.

2. *Key Results Areas* relate to the job of the individual manager. Unless it is a one-person organization, factors identified here will have some significant differences as well as some similarities with those identified under roles and missions. They fix priority on where the time, energy, and talent of the individual manager should be concentrated. Examples: Production, Cost Control, Staff Development, Customer Contact. They normally would be limited to from four to eight for each manager, concentrating on the "critical few" rather than the "trivial many."

3. *Indicators* are those factors, capable of being measured, that can be looked at within each key result area, which

will give an indication of effective or ineffective performance. Clearly, these are not absolute measurements (there
are none in management); they are capable of being
manipulated with relative ease. (I have yet to see a
management system that I couldn't "beat" if I set my mind
to it.) For them to work, those with a vested interest must
agree that the indicators selected will provide reasonable
visibility of performance. Furthermore, there must be an
assumption of integrity on the part of all concerned.
Examples: Cost Control—"unit overhead as % of direct
labor cost"; Staff Development—"number of subordinates
with a mutually agreed upon and implemented development plan." Note that "indicators" only identify *what* will
be measured, not how much or in what direction. They
serve as an intermediate step, prior to setting objectives,
designed to increase the probability that we are directing
the use of our resources to where they will get the best
payoff.

4. *Objectives* are statements of measurable results to be
achieved. Generally, they will relate to one or more of the
manager's key results areas and indicators. They can be
clearly expressed using this model: To (action verb) (single
key result) by (date) at (cost). Examples: "To reduce unit
overhead from 100% to 85% direct labor without loss of
productivity by December 31 at a cost not to exceed 60
work-hours by unit management"; "to reach agreement on
and begin implementation of an individual development
plan with not less than four of my immediate subordinates
within First Quarter at a cost not to exceed current budget
plus 40 hours of my time."

5. *Action Plans* represent the sequence of actions to be
carried out in order to achieve the objective. (This
incorporates the steps of Programming, Scheduling, and
Budgeting as described in my book.) These will be broken
down only to that amount of detail required for the
accountable manager to make his or her contribution to
the objective. Many action steps, in turn, will become

objectives for subordinates. Consequently, the responsibility for determining further detail should rest on the shoulders of the individual performing the action.

6. *Controls* are designed to keep the accountable manager informed of progress against objectives. They close the loop in the MOR process and provide the rationale for adding the phrase ". . . . and Results" to "Management by Objectives. . . ." Objectives, by themselves, are meaningless unless there is some way of ensuring their accomplishment. To the extent possible, they should be visual (I favor simple charts) and should provide for "adequate visibility in a timely fashion (sufficient to take corrective action if required) with the least expenditure of time and effort."

7. *Communication* is the catalyst that ties the whole process together. MOR is not a mechanical system, it is a human one. The process must serve as a communication vehicle among the people affected. As people become *involved* in the decisions that affect them, they become *committed* to carrying them out. That is where the real payoff comes.

As stated earlier, *Management by Objectives and Results* is a relatively simple process—deceptively simple. It is based on a common sense, logical approach to the use of familiar and proven principles and techniques of management. The difficulty comes in application which can only be tested by trying it.

# STORYBOARDING. . .
# FOR BRIEFINGS

by *Gus Matzorkis*
President
Organization Enrichment Associates
Traverse City, Michigan

Normally, you who will *give* a briefing are the same person who will *prepare* the briefing. In some special instances, you may delegate to someone else the development of a briefing plan, or the organization of briefing materials, or both. Such delegation of the actual briefing preparation is not recommended but sometimes is unavoidable.

In either event, one person normally prepares a briefing. One person, not a team.

Yet as you prepare for a briefing, you do not actually work alone or in a vacuum.

Contributions are made by others. Subordinates, boss, and/or peers provide data, ideas, and direction. This sometimes happens somewhat formally, as when you call a planning or an organizing meeting—or quite informally, as when someone tosses a suggestion or thought to you in a corridor conversation.

At that meeting or in that corridor conversation (or on the telephone, or over lunch...),you listen, you agree or disagree with what is being offered, you take notes. In the end, you:

1. Incorporate the offered help into your briefing as best you can from your notes, or

2. You consciously reject your recollection of what the offered help was, or

3. You find the offered contribution is so unclear in your mind that you use it much less effectively than you otherwise might have, or

4. You simply forget or overlook the offered help in the rush of everything else going on while you are preparing your briefing.

None of these outcomes amounts to a very fruitful utilization of the inputs and contributions of others to your briefing. Storyboarding is a better way to go. Storyboarding is particularly useful in the case of briefings which:

1. Contain technical or highly specialized information about which others in your organization are more expert than you.

2. Clearly call for new and/or innovative ways of presenting the material.

3. Will be given to an audience which is better known and understood by others in your organization than by you as the briefer.

Storyboarding methods have been used widely by the film industry, the advertising field, the aerospace industry, communications media, and by many business organizations and government agencies. Storyboarding approaches have been used in the production of films, reports, proposals, and other written documents. By now, the application of Storyboarding approaches to many group writing, group planning, and group problem solving activities is a well established success story.

The Storyboarding success story can be repeated in the preparation of your briefings as well, by replacing the casual, get-it-on-the-run manner of obtaining and using contributions from others with an organized teamwork approach to getting such contributions.

Storyboarding your briefing means:

| **This:** | **Rather Than This:** |
|---|---|
| 1. Decide who can heip in the preparation of your briefing. Get their commitment. | 1. Take what they give you if they happen to offer anything—whoever "they" turn out to be. |
| 2. Conduct a Storyboarding briefing preparation meeting, involving all committed contributors. Schedule at least 90 minutes; this is a working session. | 2. More of item 1 above. |
| 3. Use flipchart to capture ideas and suggestions of the group at the meeting. Legibility, not elegance, is the rule. Tape onto the walls the scribbled charts as they fill up with words and sketches. | 3. Make a mental note to remember what Ed mentioned to you this morning. Scrawl some hasty notes on a scratch pad as Sue is outlining a tricky point to be emphasized in the course of your briefing. |
| 4. Keep the attention at the meeting focused on the walls. As ideas and images get up there where everyone can clearly see them, people will think of new ideas and improvements for your briefing. | 4. Unable to put it off any longer, you sit down alone and begin the process of getting your briefing down on paper. |
| 5. *A few words about method:* Look for positive assertions or "Thesis Sentences" which sum- | 5. *A few words about method:* Any way that has worked pretty well for you in the past. |

| This: | Rather Than This: |
|---|---|

marize and drive home the key points in each major segment of the briefing. An effective Storyboarding technique is to tape flipchart sheets on the walls in pairs—with one Thesis Sentence for each pair, with the text or list of key supporting points on the left hand sheet, and with appropriate graphics on the right hand sheet driving home or elaborating the Thesis Sentence.

6. A good teamwork dynamic is emerging: you ask questions, get the group to engage issues, spark their thinking and yours—and all the while you write, sketch, and generally mark up (and yes, scissor and tape, too) the preliminary Storyboarding sheets which are filling up the walls.

6. You work on alone—perhaps occasionally making a phone call to ask someone a question.

7. You close the meeting. You go back over the Storyboarding sheets on the walls—revising, cleaning up, fine tuning, discarding and replacing with cleaned-up versions. You take the worked-over

7. You get things down on paper as best you can. You work in whatever way suits your style and needs in getting a typist and/or artist to produce a first draft of your briefing package.

**This:**                           **Rather Than This:**

Storyboarding sheets off
the wall and hand them
over, as appropriate, to a
typist and/or artist.

8. *Result:* A mature first cut     8. *Result:* A conventional
   at your briefing package,           first cut at your briefing
   enriched by the deliber-            package, largely based
   ately    stimulated    and         on your own thinking,
   guided inputs from con-            planning, and designing.
   tributors    during    the
   Storyboarding    meeting.

Storyboarding *is* the better way to go. It is not easy in the
sense of just doing what you normally do and casually adding a
meeting here and a little more visual approach there. There's
more to successful Storyboarding than that. You have to
develop your skills and the skills of others in doing it. Storyboarding articles, books, and seminars are three of the ways to
upgrade such skills. You have to work at Storyboarding, have to
really do it in the real world.

The nice thing is that the results will speak loudly and
clearly for themselves: better, more effective, more vividly
visual, more persuasive briefings.

# VIDEORECORDING. . .
# FOR BRIEFINGS

by *Thomas L. Sechrest*
Television Producer/Director, Division of Youth Services
Department of Health and Rehabilitative Services
State of Florida

Just as computers have become vital components of the working world, so has video technology found an increasing importance for facilitating communication in sales, staff development and training, personnel, administrative and support services, human relations and many other areas.

What exactly is video technology? It is one of the most significant developments since World War II and is truly revolutionizing old methods of communication. Videotaping involves an electromechanical procedure by which both a picture and sound are recorded on reels of magnetic tape much as a conventional tape recorded records sound only. It is today a standard and basic tool of every television station. Where it is making its most noticeable impact, however, is in its utilization in education, business and industry, the sciences and government. Here a generation of small portable videotape recorders, priced well within the reach of these consumers, has opened the door to a vast number of new ideas touching nearly every field of human endeavor—anywhere in which viewing ourselves or our activities can be useful. Videotaping provides for the first time a readily accessible and, above all, economical means of storing, retrieving and reproducing program material for instruction, documentation, research and general communication.

It is of increasing importance that those involved in *effective* business and technical presentations be aware of how the tool of nonbroadcast television can be utilized meaningfully in a variety of situations to present a well-planned, quality-controlled message to a potentially large and perhaps geographically diverse audience.

It is a communications "standard" that trainees retain about 16% of what they read, 20% of what they see, 30% of what they are told, 50% of what they both see and are told, 70% of what they see, are told and respond to, and 90% of what they see, are told, respond to and *do*! This very graphically emphasizes that making any presentation requires considerable forethought and care. A videotape program is no exception. For what purpose will the presentation be used? Will a training representative be present, or will the program be expected to relay an informative message to the viewer by itself? Will there be collateral and perhaps self-instructional materials to accompany the program? Is this indeed the best medium to use to meet an identified communications need? These are all very basic questions which must be well thought out in advance before a presentation is fully developed.

In videotaping, as in a live presentation, almost every type of auxiliary material can be adapted to and incorporated within the framework of the presentation. Charts, slides, overheads, photographs, movies, music, sound effects, dialogue, and special effects combining any or all of these materials can be effectively combined to ensure that audience interest and interpretation is at a maximal level, and the most benefit is derived from the presentation. A television lesson must, by its nature, be more tightly planned and structured than a live presentation. Its response is immediate; set up of equipment, darkening of the room and adjustment of the visual as has been a standard in the past is replaced by a lead-in and "on cue" visual which allows the instructor to present much more information in a shorter time.

Television has been used in such widespread areas as electronics training, where it overcomes a physical impossibility for a large group to easily see intricate maneuvers on specialized equipment or minute system components; in sales, where skills can be sharpened through the recording of technique and the immediate processing and critiquing by colleagues and supervisors; in personnel, where taped interviews with job candidates at conventions, college recruitment offices, etc., can reduce the time and cost of screening available applicants for positions; in public relations, where a "personal word from the President" can create and preserve a sense of organizational cameraderie; and in human relations, where group dynamics and interpersonal interactions can be studied, and patterns of communication identified and enhanced. In fact, this list is limited only by one's imagination and willingness to experiment with this training complement.

Specifically, a video recorder can be a tremendously useful tool during the "Practice" step in Preparation of a Briefing. The opportunity to both see and hear oneself while making a presentation is a revealing and sometimes traumatic experience. However, in terms of impact and ultimate learning, there is probably no better teacher. In a program designed to train or coach individuals in the making of a briefing, our recommendation would be that the playback be done in a private session with the briefer and a trainer/coach. In this way, it can be stopped, re-run, discussed and, where necessary, a plan for correction can be developed without the briefer feeling as though he or she were in a goldfish bowl.

With the rapid expansion of the use of videotape in training and general communication, many specialists have developed programs available for rental, lease, or purchase by those firms or agencies who do not have the facilities for the development of their own programs. Depending on individual and particular communication needs, the utilization of these productions may or may not prove adequate. Previewing available materials allows you to estimate effectiveness versus cost, and may provide the necessary impetus or rationale for the "big step" of outfitting a complete in-house studio.

Development of a presentation capability with videotape will afford the user several distinct advantages.

1. Instructors who are recognized in the field can be utilized to add interest and credibility to the endeavor.

2. Adaptable course materials can be provided to supplement the televised lesson and provide written practice and measurable followup on course content.

3. Designated group leaders within the organization can facilitate a discussion session immediately following the televised lesson which is based on the lesson seen and which relates the content of the program and the supplementary materials to the needs of the discussion participants and their activities within the organization.

4. Perhaps most convenient, however, is the ability to schedule programming at the most convenient and useful times, or to replay lessons on an individual or collective basis so that comprehension of content is guaranteed.

5. An added plus is the more than reasonable cost, which never fails as a demonstrable selling point.

Whether in formal educational institutions or in the classrooms of business, industry and government, the new medium of nonbroadcast television has reached a level of practical maturity. Daily, the videotape program is expanding its advantages of communication into classrooms, board rooms, offices, factories, and even employee lounges. Whether the subject is supervision, management by objectives, secretarial skills, statistics, business writing, job enrichment or verbal communication, television is able to provide the means of getting more information to more people more effectively. The fact is, television is here to stay, and the sooner you investigate its applicability to the presentations you make, the better your chances will be of providing low cost and efficient programming to meet the ever-increasing demands placed on you by the need for effective communication.

# APPENDIX B

**AUDIENCE ANALYSIS AUDIT (AAA)\***
(Fill in the blanks or circle the terms most descriptive)

---

1. Identify the objectives in presenting your briefing to THIS audience. What do you want to happen as a result of it?

   _____

   _____

   (Keep these objectives in mind as you consider the items below.)

2. **Specific Analysis** of members of this audience —
   a. Their knowledge of the subject:

      High level     General    Limited    None    Unknown
   b. Their opinions about the subject and/or the speaker or organization represented:

      Very favorable      Favorable         Neutral
      Slightly hostile      Very Hostile      Unknown
   c. Their reasons for attending this briefing:

      _____

   d. Advantages and disadvantages of briefing objectives to them as individuals:

      Advantages _____

      Disadvantages _____

3. **General Analysis** of members of this audience —
   a. Their occupational relationships to the speaker or organization:

      Customer    Top management    Immediate management
      Peers        Subordinates      Other management
      Other workers    Public

---

*The material on these perforated pages may be reproduced for the convenience of the reader.

**Audience Analysis Audit (AAA) (continued)**

---

   b. Length of relationship with organization as customer or employee:

     New     Less than two years     More than two years
     Unknown

   c. Their vocabulary understanding level:

     Technical          Nontechnical     Generally high
     Generally low      Unknown

   d. Open-mindedness (willingness to accept ideas to be presented)

     Eager          Open          Neutral
     Slightly resistant     Strongly resistant     Unknown

4. Information and Techniques —

   a. Information and techniques most likely to gain the attention of this audience:

     Highly technical information     Statistical comparisons
     Cost figures     Anecdotes     Demonstrations     Other

     _____

   b. Information or techniques likely to get negative reactions from this audience:

     _____

5. Summarize, in a few sentences, the most important information from the preceding four sections.

     _____

     _____

     _____

     _____

     _____

     _____

## GUIDELINES FOR PREPARING A PRELIMINARY PLAN*

1. Identify specific objectives for the briefing, keeping in mind the following criteria:
   a. They should answer the question, "Why am I giving this briefing?"
   b. They should state the results desired from the briefing, in effect, completing the sentence, "I want the following things to happen as a result of this briefing: . . ."
   c. They should be designed to accomplish whatever hidden objectives you have for the briefing.

Note: If the body of knowledge to be presented must be identified in the objectives, use a sentence such as "I want to tell about . . . so that . . . will take place."

2. Identify the specific audience for whom you are designing this briefing and state in a one- or two-sentence summary pertinent information about their knowledge, attitudes, and so forth.

---

*The material on these perforated pages may be reproduced for the convenience of the reader.

## Guidelines for Preparing a Preliminary Plan (continued)

---

3. State the MAIN IDEAS OR CONCEPTS that the audience MUST get if the objectives of the briefing are to be met. These should:
   a. Be in conclusion form and preferably in complete sentences.
   b. Definitely lead to the accomplishment of the specific objectives.
   c. Be interesting in themselves or capable of being made so.
   d. Be few in number, usually no more than five.

4. Identify under each main idea the types of factual information necessary so that this audience can understand these ideas. Avoid excessive detail.

### This Plan Should be Used as a Guide:

1. For the briefer in selecting materials, keeping ideas channeled, and determining emphasis points.

2. For support personnel who may provide the backup data, prepare charts and other aids, and assist in the briefing itself.

**PRELIMINARY PLAN WORKSHEET\***

Title or subject of this briefing : _____

_____

Approximate date, time, and place of this briefing : _____

_____

Who requested the briefing (if other than yourself)? _____

_____

Your **OBJECTIVES** for *this* briefing (what will be the immediate results if this briefing is successful?) :

1. _____

   _____

2. _____

   _____

3. _____

   _____

4. _____

   _____

**AUDIENCE** for this briefing (who are they and what is their general knowledge of, interest in, and attitude toward the subject?) :

_____

_____

_____

---

\*The material on these perforated pages may be reproduced for the convenience of the reader.

**Preliminary Plan Worksheet (continued)**

---

**MAIN IDEAS OR CONCEPTS** that the audience *must get and retain* if the objectives of the briefing are to be met:

1. _____

_____

2. _____

_____

3. _____

_____

4. _____

_____

5. _____

_____

Types of **FACTUAL INFORMATION** necessary to support the main ideas:

Idea 1

_____

_____

_____

_____

Idea 2

_____

_____

_____

_____

**Preliminary Plan Worksheet (continued)**

---

Idea 3

_____

_____

_____

_____

Idea 4

_____

_____

_____

_____

Idea 5

_____

_____

_____

_____

## RESOURCE MATERIAL SELECTION WORKSHEET*

Title or Subject: _____

Briefer(s): _____

Time and Place: _____

Date: _____

1. What is the *object or purpose* of the briefing?
   a. Parts of Preliminary Plan to consider _____

   b. Specific reference material needed _____

   _____

   _____

2. What should be *covered*? What can best be *eliminated*?
   a. Parts of Preliminary Plan to consider _____

   b. Specific reference material needed _____

   _____

   _____

3. What amount of *detail* is necessary?
   a. Parts of Preliminary Plan to consider _____

   b. Specific reference material needed _____

   _____

   _____

4. What *must* be said if the objectives are to be reached?
   a. Parts of Preliminary Plan to consider _____

_____

*The material on these perforated pages may be reproduced for the convenience of the reader.

**Resource Material Selection Worksheet (continued)**

---

    b.  Specific reference material needed _____

_____

_____


5.  What is the *best* way to say it?
    a.  Parts of Preliminary Plan to consider _____

    b.  Specific reference material needed _____

_____

_____


6.  What kind of audience *action or response* is required if the objectives are to be met?
    a.  Parts of Preliminary Plan to consider _____

    b.  Specific reference material needed _____

_____

_____


7.  What material should be *withheld* from the briefing itself, but be available for reference during the question-and-answer period?
    a.  Parts of Preliminary Plan to consider _____

    b.  Specific reference material needed _____

_____

_____


8.  Finally, submit all resource material to the *"Why?"* test.

_____

## PRESENTATION WORKSHEET*

Title or Subject : _____

Briefer(s) : _____

Date, Time and Place : _____

### General Considerations

1. How will the room be arranged (seating arrangements, lighting, namecards, etc.)? _____

    _____

    _____

2. How and when will the audience be notified of the briefing? Approximately how many will attend? _____

    _____

    _____

3. What equipment, aids, and supplies will be required? How will they be transported to the briefing location? _____

    _____

    _____

4. What handout materials will be required? What arrangements have to be made for them? How and when will they be distributed at the briefing? _____

    _____

    _____

5. How and when will you handle audience questions? _____

    _____

    _____

_____

*The material on these perforated pages may be reproduced for the convenience of the reader.

**Presentation Worksheet (continued)**

---

**Presentation Outline**

| Time | Content | Methods/<br>aids/examples |
|------|---------|----------------|
| | STATE THE IDEA (Introduction)<br>(Sell the audience on listening;<br>introduce the subject)<br><br>DEVELOP THE IDEA (Body)<br><br>RESTATE THE IDEA (Conclusion) | |

**Presentation Worksheet (continued)**

**Presentation Outline**

| Time | Content | Methods/<br>aids/examples |
|------|---------|---------------------------|
|      |         |                           |

**BRIEFING EVALUATION GUIDE***

---

Presentor: _____ Evaluator: _____

## CONTENT

### Introduction

1. How good was it in arousing interest in the briefing?
   Outstanding_____Good_____Fair_____Weak_____

2. Was the purpose of the briefing made clear?
   Yes_____Somewhat_____No_____(?)†_____

Comments:

### Body

1. Did the main ideas come through clearly?
   Yes_____Somewhat_____No_____(?)_____

2. Were the supporting ideas and illustrations:
   Interesting?        Yes_____Somewhat_____No_____
   Varied?             Yes_____Somewhat_____No_____
   Directly related?   Yes_____Somewhat_____No_____

---

*The material on these perforated pages may be reproduced for the convenience of the reader.

†The items marked "?" are for those instances where the evaluator does not consider himself technically competent to judge and says, in effect, "I don't know."

**Briefing Evaluation Guide (continued)**

---

3. Was the presentation appropriate for the audience (as identified by briefer)?
Yes_____Reasonably so_____No_____(?)_____

Comments:

**Conclusion**

1. Did it sum up main ideas and purposes?
Yes____Somewhat____No_____(?)_____

2. How effective was it in encouraging action, belief, understanding?
Outstanding_____Good_____Fair_____Weak_____

Comments:

**General**

1. How would you grade the briefing?
Outstanding_____Good_____Fair_____Weak_____

2. Were the objectives of the briefing likely to be reached?
Yes_____Probably_____No_____(?)_____

Comments:

**Briefing Evaluation Guide (continued)**

---

## PRESENTATION

### Audio-Visual Aids

1. Were they suited to the topic and the audience?
   Yes_____Reasonably so_____No_____

2. Would they be visible to everyone and easy to follow?
   Yes_____Reasonably so_____No_____

3. How effective was the use of these aids?
   Outstanding_____Good_____Fair_____Weak_____

Comments:


### Platform Techniques

1. Poise: Was the presentor in control of the situation?
   Yes_____Reasonably so_____No_____

2. Were posture and movements appropriate?
   Yes_____Reasonably so_____No_____

3. Were gestures effective?
   Good_____Fair_____Overdone_____Ineffective_____

4. Was relationship with the audience effective (eye contact, etc.)?
   Outstanding_____Good_____Fair_____Weak_____

Comments:

**Briefing Evaluation Guide (continued)**

---

**Vocal Techniques** (Check more than one, if necessary)

1. How about pitch and quality?
   Good_____Too high_____Too low_____
   Monotonous_____Harsh_____Nasal_____

2. How about rate and intensity?
   Good_____Too fast_____Too slow_____Too loud_____
   Too soft_____Monotonous____

3. Did he or she speak clearly and distinctly?
   Yes_____Reasonably so_____No_____

Comments:


**General**

1. How did you feel about the speaker's overall presentation?
   Outstanding_____Good_____Fair_____Weak_____

2. Make any general comments you feel would be helpful.

## EFFECTIVE AIDS TO UNDERSTANDING*

The following aids can be used to make the subject easier to understand or more interesting, and to promote the kind of thinking that will help you accomplish your objectives:

### 1. Charts

To direct thinking; clarify a specific point; summarize; show trends, relationships, and comparisons.

Information charts or tabulations should usually be prepared in advance to ensure that all points are covered and covered accurately.

### Types of charts

**Highlights**, straight copy, emphasizing key points.

**Time-Sequence** (historical), showing relationships over a period of time. May be in seconds or centuries. Can use pictures or graphs.

**Organizational,** indicating relationships between individuals, departments, sections, or jobs.

**Cause-and-Effect,** e.g., picture of bottle plus driver plus auto equals wrecked auto.

**Flow Chart,** to show relations of parts to finished whole or to the direction of movement. A PERT (Program Evaluation and Review Technique) chart is a flow chart.

**Inventory,** showing picture of object and identifying parts off to the side by arrows.

---

*The material on these perforated pages may be reproduced for the convenience of the reader.

**Dissection,** enlarged, transparent, or cut-away views of object.

**Diagrammatic or Schematic,** reducing complex natural objects by means of symbols to simple portrayal, e.g., radio wiring diagram.

**Multi-Bar Graph,** using horizontal or vertical bars representing comparable items.

**Divided-Bar Graph,** a single bar divided into parts by lines to show the relation of parts to the whole.

**Line Graph,** using a horizontal scale (abscissa) and vertical scale (ordinate), e.g., showing number of overtime hours being worked each month.

**Divided Circle,** pie graph, used in the same way as the divided bar.

**Pictograph,** a pictorial symbol representing comparable quantities of a given item, e.g., stacks of coins representing comparable costs of different phases of an operation.

## 2. Illustrations, Diagrams, and Maps

To clarify a point, emphasize trends, get attention, or show relationships or differences.

## 3. Motion Pictures

To show motion, give overall view or impression, or show actual operation.

## 4. Film Strips or Slides

Same as for motion pictures. They are especially helpful when motion is not required or desirable. They are not expensive and can be studied as still pictures.

## 5. Samples or Specimens

To show the real object.

## 6. Models

*Small-scale*—To permit showing an operation without using large quantities of material, to make a large operation visible, or to show a project to be completed.

*Large-scale*—To make an object large enough to permit handling, identify small parts, or see internal operation.

## 7. Exhibits

To show finished products, the results of good and poor practices, attract attention, arouse and hold interest, and adequately display one idea. (Use life, motion, color, or light to help attract attention.)

## 8. Worksheets

To provide "hands on" experience by audience in performing certain actions; to provide a carry over to the job.

## 9. Manuals, Pamphlets, Instruction Sheets, Circular Letters, Outlines and Bulletins

For standard information and guides, for reference and background.

## 10. Cartoons, Posters, Signs

To attract attention and arouse interest.

## 11. Photographs, Textbook or Magazine Illustrations

To tie discussion to actual situations and people, provide current interest, or show local activities.

## 12. Case Studies

To bring together for specific situations, the principles, practices, and procedures which are being explained, interpreted, or formulated by the group. It is much easier to visualized a procedure if you "Take the case of Mr..."

## 13. Examples and Stories

To break monotony or tension, fix an idea, get attention, illustrate or emphasize a point, clarify a situation, or break away from a delicate or ticklish subject.

## 14. Demonstration

To show how to carry out a suggested method or procedure.

## 15. Field Trips

To present subject in its natural setting, stimulate interest, blend theory with practicality, and provide additional material for study.

## PRELIMINARY ARRANGEMENTS CHECKLIST*

Session:_____Briefer:_____ Coordinator:_____

Date & Time:_____Room: _____ No. Expected:_____

**Seating arrangement** (No. per table_____)

Auditorium_____Classroom_____Informal_____U-Shape_____

Buzz_____      Herringbone_____Other_____

**Room considerations**

|  | Satis-factory | Not Needed | Needs Attention | Comments |
|---|---|---|---|---|
| Chairs | | | | |
| Tables | | | | |
| Lighting | | | | |
| Ventilation | | | | |
| Distractions | | | | |
| Ash trays | | | | |
| Pencils/scratch paper | | | | |
| Name-cards | | | | |
| Coffee/soft drinks/water | | | | |
| | | | | |
| | | | | |
| | | | | |

*The material on these perforated pages may be reproduced for the convenience of the reader.

**Preliminary Arrangements Checklist (continued)**

## Equipment/aids/handouts

| (List) | Satis-factory | Not Needed | Needs Attention | Comments |
|---|---|---|---|---|
|  |  |  |  |  |
|  |  |  |  |  |
|  |  |  |  |  |
| Electrical accessories (bulb, cord, plug, extension) |  |  |  |  |
|  |  |  |  |  |
|  |  |  |  |  |
|  |  |  |  |  |
| Supplies (chalk, eraser, felt pens, grease pencil, tape) |  |  |  |  |
|  |  |  |  |  |
|  |  |  |  |  |
| **Audience notification** |  |  |  |  |
|  |  |  |  |  |
|  |  |  |  |  |